Introduction to **The Physics of Rocks**

Il ne se laissait pas de soupeser et d'étudier curieusement les pierres dont les contours polis ou rugueux, les tons de rouille ou de moisissure racontent une histoire, témoignent de métaux qui les ont formées, des feux ou des eaux qui ont jadis précipité leur matière ou coagulé leur forme.

—Marguerite Yourcenar, *L'Oeuvre au Noir*

For hours at a time he would examine stones, weighing them and studying their rough or polished contours, their coloration from rust or mold, all of which tell a tale and testify as to the metals which have composed them, the waters which long ago precipitated their substance, and the fires which have coagulated them into the shapes we see.

—Marguerite Yourcenar, *The Abyss*

Introduction to **The Physics of Rocks**

Yves Guéguen and
Victor Palciauskas

PRINCETON UNIVERSITY PRESS · PRINCETON, NEW JERSEY

Copyright © 1994 by Princeton University Press
Published by Princeton University Press, 41 William Street,
Princeton, New Jersey 08540
In the United Kingdom: Princeton University Press, Chichester, West Sussex

Library of Congress Cataloging-in-Publication Data

Guéguen, Yves.
[Physique des roches. English]
Introduction to the physics of rocks/ Yves Guéguen
    and Victor Palciauskas.
        p. cm.
Includes bibliographical references and index.
ISBN 0-691-03452-4
1. Petrology. 2. Rock mechanics. I. Palciauskas, Victor, 1941-.
II. Title.
    QE431.5.G8413   1994
    552'.06—dc20   93-40793 CIP

Excerpt from *The Abyss* by Marguerite Yourcenar, translated by Grace Frick.
Translation copyright ©1976 by Marguerite Yourcenar. Reprinted by permission of
Farrar, Straus & Giroux, Inc.

This book has been composed in Times Roman

Princeton University Press books are printed on acid-free paper, and meet the guidelines for
permanence and durability of the Committee on Production Guidelines for Book Longevity of
the Council on Library Resources

Printed in the United States of America

10 9 8 7 6 5 4 3 2 1

Contents

Preface and Acknowledgments

THE UNPRECEDENTED increase in both the quality and quantity of geophysical data has dramatically improved our knowledge of the earth's subsurface. This development was motivated not only by increased exploration for various mineral and energy resources (gas, oil, ores, geothermal energy) but also by the study of natural and man-made hazards (earthquakes, nuclear waste storage, etc.). Acoustic, electric, and other measurements are now routinely obtained from well logs and large-scale geophysical surveys. The number of parameters measured and the accuracy of measurement are continuously increasing. To translate these measurements into rock properties and other information of economic interest is the realm of *rock physics*. Rocks are inherently complex, inhomogeneous systems. Inferring lithology, porosity, and fluid content from geophysical signals or modeling the response of a medium with ever-increasing accuracy requires a solid physical understanding of these systems. For these reasons exploration companies, universities, and government research laboratories are now recognizing the important role of this field of study.

This book has been written for geologists, geophysicists, and other earth scientists. It is intended to be an introduction to the physics of rocks and does not pretend to be exhaustive. Rather, its aim is to provide a basis for understanding the various rock properties and how these properties depend on the rock type and its history. One of the principle objectives is to provide geologists and geophysicists with a common physical basis for communication. Geologists know rocks very well, but often they shy away from mathematical presentations concerning the physical properties. The reverse can also be true: Geophysicists do not always know the history of the rocks on which they have accumulated substantial data.

The first three chapters are a general introduction to rocks and how their microstructure reflects their origin and evolution. We shall present the recent advances in modern physics which allow us to study rocks as porous and inhomogeneous media. Empirical laws, which were introduced long ago to model the complexity of rocks, now appear to be of a more general nature and in line with fundamental physical principles. The eight following chapters deal with the different physical properties of rocks. Mechanical properties are considered first. Next, the important role of fluids in porous materials is emphasized, followed by a chapter each on acoustic, electrical, dielectric, thermal, and magnetic properties.

This book has greatly benefited from discussions with many colleagues. We wish to thank them all, with specific thanks to M. Darot and D. Jeannette for their many contributions. We have gained immeasurably from the continuous interaction with young research fellows: among them, C. David, P. Gavrilenko, T. Reuschle, C. Ruffet. We also thank those who led us to discover new theoretical concepts and experimental results: Y. Bernabe, T. Chelidze, J. Dienes, J. P. Gratier, P. Meredith, J. Schopper, and L. P. Stephenson. Many colleagues allowed us to reproduce their original figures from their publications. We thank in particular R. Lenormand and E. Pittmann. We also thank M. Willer for her long and patient typing and S. and H. Paff for the illustrations.

This book would not have been possible without the support and patience of Christine and Aurelia. In order to thank them, we include them in our common goal. We hope that this book will be a useful and efficient tool for students and researchers. We hope that it will contribute to a better understanding of our planet and to a better management of its resources.

Introduction to **The Physics of Rocks**

I. Rocks

MINERALS AND ROCKS

Minerals

Minerals are the naturally occurring chemical compounds of the earth. They are natural examples of solid phases of matter that are homogeneous and are separated from other substances by well-defined boundaries. Minerals are usually found as grains whose diameters range in size from submicroscopic to a few centimeters. To define a mineral uniquely, both the chemical composition and molecular structure must be specified. For example, the molecular formula SiO_2 is common to the minerals α-quartz, β-quartz, opal, cristobalite, tridymite, coesite, and stishovite: the differences being in the periodic, three-dimensional arrangement of oxygen and silicon atoms. Minerals which have the same chemical composition but different structures are called *polymorphs*. They can pass from one structure to another when the temperature and pressure conditions are changed, and one phase becomes more stable than the other. Although in most cases individual grains are crystalline, one also finds amorphous minerals which are usually formed from rapidly cooling silicate melts. Because each mineral possesses a specific chemical composition and molecular structure, it will also posses a distinct set of physical properties. These properties characterize how the mineral will respond to external disturbances such as mechanical forces, electric and magnetic fields, and temperature changes. The measurement and prediction of mineral properties has been a primary concern of solid-state physics for the past forty years, and thus a wide body of knowledge has been accumulated in this domain.

Rocks

Just as minerals are aggregates of atoms and can exhibit properties not possessed by any individual atom, rocks are mineral aggregates and can exhibit properties that are not possessed by any individual mineral. The physical, chemical, and geometric properties of rocks depend on the physical, chemical, and geometric properties of the individual minerals, their volume fractions, and their distribution. *Microstructure* is a term used to denote this complex internal geometry of a rock. Rock microstructures exhibit a wide variety of inhomogeneities (disorders): disorder in mineral arrangements, variability in mineralogy and grain size, number and sizes of pores, degree of fracturing, etc. One fascinating characteristic of rocks is a direct consequence of this wide range of possible inhomogeneities: the physical properties of a rock can depend on the scale of measurement. Consider for example an idealized rock composed of a random aggregate of identical grains. In this case only one type of disorder is apparent, that being the lack of long-range order in grain packing. This particular inhomogeneity or disorder can be observed only at the grain scale. When viewed on a scale much larger than the grain size, the rock appears uniform and can be considered statistically homogeneous. That is, on a sufficiently large scale all parts of the rock have similar physical properties that are scale independent. Rocks found in nature rarely exhibit such simple behavior and usually show some variation in properties as the scale of measurement changes. This phenomena results from the fact that inhomogeneities in nature arise from a great variety of sources acting at many length scales. Some common examples are finite size and packing of mineral grains, distribution of grain sizes, variety of minerals, spatial distribution of sizes (sorting), and rock fracture. All of these inhomogeneities reflect the dynamics of sediment formation and subsequent evolutionary changes. It is often found that the scale of the inhomogeneities increases with the size of the rock considered and that there is no practical length scale above which the rock properties are scale independent. This poses a fundamental problem that is not encountered in mineral physics. How do we extrapolate laboratory-measured rock properties which are determined on centimeter-sized samples, to field-scale problems measured in kilometers? The aim of this book is to describe this rich behavior in rock properties and show how these properties can be related to the constituent minerals and the internal rock microstructure.

Classification of Rocks

A complete description of rock microstructure in terms of constituent minerals, their volume fractions, spatial distribution, etc. is usually impractical, if not impossible. For this reason various classification schemes have been devised which convey most of the information about a rock in a simpler, but less precise, manner. Because rocks have a wide range of properties, all classification schemes must focus on only two or three relevant properties. The choice of properties is usually dictated by the application. For example, when considering fluid transport through a rock, the magnetic susceptibility of the mineral grains will be less important than the mineral grain size. Thus a classification scheme based on grain size would be relevant for this purpose.

One of the most popular methods of rock classification has been in terms of the physical processes responsible for their origin. Three major categories of processes have been identified:

Igneous Processes: crystallization of minerals and solidification of glasses (amorphous silicates) from high-temperature silicate melts.

Sedimentary Processes: weathering of preexisting rocks (sedimentary, igneous, and metamorphic), transport of weathered products by such media as wind and moving water, and deposition of suspended material from air or water. Sedimentary rocks are also formed through chemical processes such as solution and precipitation of minerals by water and secretion of dissolved materials through organic agents.

Metamorphic and Diagenetic Processes: during burial and heating, rocks can experience recrystallization and mutual reaction of constituent minerals as their stability fields are exceeded. Because these reactions take place without ever reaching the silica melt phase, they are called metamorphic. Pore fluids such as water also play a very important role in the evolution of a rock and its microstructure. Exchange of ions between minerals and moving fluids, the dissolution and precipitation of minerals (cements), or simply the long-range transport of mass can dramatically alter the composition and physical properties of a rock.

After formation most rocks are exposed to a series of processes and cannot be classified by a single process. This book will not focus on the detailed geologic history when describing a rock, but rather on its present mineralogy and microstructure and how they affect the measurable rock properties. For this reason it will be sufficient to classify rocks into two major categories, igneous and sedimentary. Metamorphic rocks will not be viewed as a separate rock type but as evolutionary stages of sedimentary rocks. In many cases it is possible to predict, at least semiquantitatively, how a diagenetic-metamorphic process will alter microstructure. By developing an understanding of how microstructure affects physical properties, an understanding of how diagenetic processes affect physical properties can be gained. For a more detailed description of rocks, their geologic history, and their chemical and physical evolution, we recommend the books cited at the end of this chapter.

IGNEOUS ROCKS

Igneous rocks are considered to be the result of solidification of molten material that originated in the Earth's interior. Magmas which have flowed out onto the earth's surface and cooled rapidly form volcanic (or extrusive) rocks. Those that do not reach the surface and solidify slowly in the subsurface form plutonic (or intrusive) rocks. Mineralogically there is a general similarity between corresponding extrusive and intrusive rocks: the principal constituents being silicates. The major differences can be attributed to the rapid cooling under volcanic conditions which permit the formation of metastable phases, such as silicate glasses. Metastable phases are generally not found in plutonic rocks.

Chemical Composition

Igneous rocks can be subdivided into three groups based on chemical composition: acidic, intermediate, and basic (and ultrabasic) rocks. A rock rich in silicate (SiO_2) is called "acidic" while a rock less rich in SiO_2 is called "basic"; the range being $45\% < SiO_2 < 75\%$. Acidic rocks contain sufficient silica for the mineral quartz to be present (about 10%). Basic rocks, on the other hand, do not have sufficient silica to have quartz present. Less silica is found in feldspars, which contain
other cations: Al, Na, K, Ca (table I.1). Other elements, Mg and Fe in particular, are required for olivines, pyroxene, and amphiboles (table I.1). By simplifying to three types, igneous rocks can be classified by chemical composition (fig. I.1). The classification introduced corre-

Table I.1. MAJOR MINERALS OF IGNEOUS ROCKS

Mineral	Composition	Crystal System	Formula Weight	Density (g / cm^3)
Quartz	SiO_2	Hexagonal	60.1	2.65
Feldspar				
Orthoclase	$KAlSi_3O_8$	Monoclinic	278.4	2.55
Plagioclase				
Albite	$NaAlSi_3O_8$	Triclinic	262.2	2.62
Anorthite	$CaAl_2Si_2O_8$	Triclinic	278.2	2.76
Olivine				
Forsterite	Mg_2SiO_4	Orthorhombic	140.7	3.21
Fayalite	Fe_2SiO_4	Orthorhombic	203.8	4.39
Pyroxene				
Enstatite	$MgSiO_3$	Orthorhombic	100.4	3.12
Diopside	$CaMg(SiO_3)_2$	Monoclinic	216.6	3.28
Hedenbergite	$CaFe(SiO_3)_2$	Monoclinic	248.1	3.55
Amphibole				
Tremolite	$Ca_2Mg_5Si_8O_{22}(OH)_2$	Monoclinic	812.5	2.98
Iron tremolite	$Ca_2Fe_5Si_8O_{22}(OH)_2$	Monoclinic	970.1	3.40
Glaucophane	$Na_2Mg_3Al_2Si_8O_{22}(OH)_2$	Monoclinic	783.6	2.91

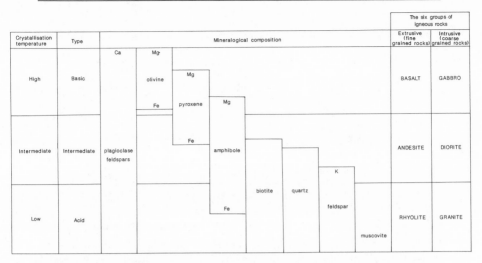

Fig. I.1. The six groups of igneous rocks associated with the three principal groups: basic, intermediate, and acidic. (After *Cambridge Encyclopedia of Earth Sciences*, 1981.)

sponds to a decreasing temperature of crystallization: for $CaAl_2Si_2O_8$, $T_f = 1550°C$ while for $NaAlSi_3O_8$, $T_f = 1100°C$ (at $P = 1$ atmosphere). From the results in fig. I.1 it can be seen that certain minerals are frequently found together: for example, olivine, pyroxene, and calcium plagioclase. Others, such as quartz and pyroxenes, never appear together. Bowen was the first to point out the significance of these features and provide the most convincing evidence supporting the igneous origin of plutonic rocks.

There exists an approximate inverse correlation between the temperature at which a mineral crystallizes from a magma and its relative stability to alteration processes affecting all sedimentary rocks. Olivine and pyroxenes, for example, are minerals formed at high temperatures and are easily altered. At the other extreme, quartz resists most alteration. It is possible to calculate the "lifetime" τ of a mineral grain, 1 mm in diameter, at given pH and temperature conditions. The calculated lifetimes of several minerals at pH = 5, $T = 25°C$ conditions are shown in table I.2.

Microstructure

Microstructure introduces a second subdivision of igneous rocks. One can distinguish whether a rock is composed of large or small grains: granite and rhyolite, diorite and andesite, gabbro and basalt. In general the first are intrusive rocks (slow cooling) and the second extrusive

Table I.2. LIFETIME OF MINERAL GRAINS (1 MM DIAMETER)

Mineral	τ (years)
quartz	$34 \cdot 10^6$
muscovite	$3 \cdot 10^6$
forsterite	$600 \cdot 10^3$
K-feldspar	$520 \cdot 10^3$
albite	$80 \cdot 10^3$
enstatite	$8 \cdot 10^3$
anorthite	100

Source: After Lasaga, A. C., 1984, Chemical kinetics of water-rock interactions, *Journal of Geophysical Research* 89:4009–25.

(rapid cooling at the surface). The classification of igneous rocks into six groups, based on silica content and grain size, is illustrated in figure I.1 and is capable of accounting for 90% of igneous rocks.

Rock microstructure is strongly related to its origin. Basaltic magmas contain much dissolved gas, which forms enclosed bubbles resulting in a porous structure. A "porphyritic" texture (numerous small crystals enclosing larger preexisting crystals) arises due to the rapid cooling. Granites on the other hand (fig. I.2) have a structure composed of large crystals and almost no porosity: their origin being a slow cooling at elevated pressure. The later history of an igneous rock can greatly modify this initial microstructure (later fracturing, etc.).

Fig. I.2. Thin section of granite (Auriat granite) photographed with a polarization microscope: Muscovite in the interior of feldspar. Two microfissures traverse the muscovite grain. The small line represents 0.7 mm.

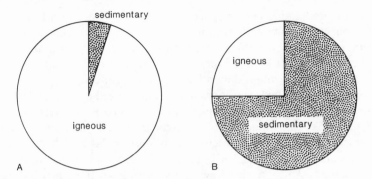

Fig. I.3. Relative abundances of igneous and sedimentary rocks in the crust. *A*: by volume; *B*: by surface. (After Pettijohn, 1975, p. 19.)

SEDIMENTARY ROCKS

Although igneous rocks represent 95% of the earth's crust by volume, approximately 75% of the continental surface and almost the totality of the ocean floor is covered by sedimentary rocks (fig. I.3). Sedimentary rocks are the result of desegregation of igneous rocks by a series of physical-chemical processes, where the weathering agents wind and water play a major role. The thickness of the sedimentary cover averages 2 km over continental areas and approximately 1 km on the ocean floors (Pettijohn, 1975).

Mode of formation

The microstructure of clastic sediments is due to their mode of formation. Most igneous rocks crystallize in place, and thus the constituent minerals are in continuous and tightly interlocking contact with each other. On the other hand, each clastic particle forming a sedimentary rock (sand grain, pebble, shell fragment, etc.) has been formed elsewhere, transported, and then placed in the present location within the rock. For this reason, at formation the individual grains are not in continuous contact with each other and a significant void space free of solids is enclosed within the rock. This formation process also results in large stress inhomogeneities within the rock. Because the weight of the overlying sediments is supported by a relatively small contact area between grains, there is a large stress concentration at grain-to-grain contacts. Under these conditions, various chemical and physical processes are activated whose role is to diminish these stress inhomogeneities by deforming the grains, increasing the contact area, and decreasing the pore space.

Young fine-grained oceanic sediments can have porosities as large as 80%, but at depth porosities range 5–30% on the average. Mobile fluids in the pore space constitute a medium for numerous chemical reactions and mass transport between minerals of the host rock. In addition these porous and permeable sediments can be the location of large accumulations of trapped hydrocarbons. Thus, understanding the evolution of sedimentary rocks and their physical properties is of great economic value. Unlike the minerals of igneous rocks, the minerals of clastic sedimentary rocks are not equilibrium assemblages. They were not precipitated in equilibrium with each other or with the fluid from which they settled. Due to the low temperatures at the earth's surface, large-scale mineral reactions and transformations do not take place until the sediments are buried and the temperature and pressure are increased. Then various "diagenetic" reactions take place altering the rock composition and microstructure. Rock evolution is controlled by the strong coupling between fluid circulation, chemical reaction, and pore space geometry.

The Four Principal Groups

Sedimentary rocks have been grouped into four main categories: sandstone, shale, carbonate, and evaporite. This group classification focuses partly on the size of the sedimentary particles (sandstone and shale) and partly on mineralogy and process of formation (carbonate and evaporite).

(*a*) *Sandstones* are defined as rocks that are composed of particles ranging in size from 1/16 to 2 mm in diameter. They are generally composed of individual mineral grains or rock fragments that have been derived from the weathering of crystalline rocks, such as granite or preexisting sandstones. Their microstructure reflects their origin: grains are rounded by transport, sorted in size by the depositional environment, unstable minerals are eliminated after long alteration, etc. Thus sandstones tend to be accumulations of minerals that are chemically and physically durable. Quartz is the major constituent with feldspars, especially K-feldspar, being quite common. Segregation of quartz into the larger grain size sediments is an example of "sedimentary differentiation." When "sedimentary differentiation" is very strong, one obtains a very clean sandstone, composed of 100% quartz (Fontainebleau sandstone, fig. I.4). The average mineral composition of a sandstone is shown in figure I.5. Sandstones represent approximately 25% of sedimentary rocks and are of great economic significance. They are important natural reservoirs for gas, oil, and water. They have a variety of industrial uses: as abrasives; as raw

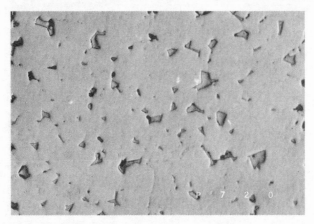

Fig. I.4. Thin section of a sandstone (Fontainebleau sandstone) photographed by a reflection microscope. The visible pores are equal and situated between points of contact of quartz grains. The small line represents 1.5 mm.

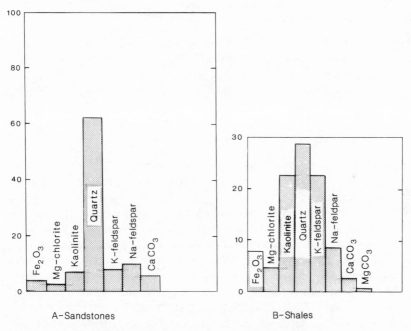

Fig. I.5. Average chemical and mineralogical composition of sandstone and shale. (After Garrels and MacKenzie, 1971, pp. 206–10.)

Table I.3. MAJOR MINERALS OF SEDIMENTARY ROCKS

Mineral	Composition	Crystal System	Formula Weight	Density (g/cm^3)
Quartz	SiO_2	Hexagonal	60.1	2.65
Feldspar				
Orthoclase	$KAlSi_3O_8$	Monoclinic	278.4	2.55
Plagioclase				
Albite	$NaAlSi_3O_8$	Triclinic	262.2	2.62
Anorthite	$CaAl_2Si_2O_8$	Triclinic	278.2	2.76
Phyllites				
Chlorite	$Mg_5Al_2Si_3O_{10}(OH)_8$	Monoclinic	555.7	2.6–2.8
Montmorillonite	$Na_{0.33}Al_{2.33}Si_{3.67}O_{10}(OH)_2$	Monoclinic	257.5	2–3
	$Ca_{0.17}Al_{2.33}Si_{3.67}O_{10}(OH)_2$	Monoclinic	256.6	Variable
Muscovite (mica)	$KAl_3Si_3O_{10}(OH)_2$	Monoclinic	398.3	2.8
Kaolinite	$Al_2Si_2O_5(OH)_4$	Monoclinic	258.1	2.65
Carbonates				
Calcite	$CaCO_3$	Hexagonal	100.1	2.71
Aragonite	$CaCO_3$	Orthorhombic	100.1	2.93
Dolomite	$CaMg(CO_3)_2$	Hexagonal	184.4	2.87
Magnesite	$MgCO_3$	Hexagonal	84.3	3.01
Evaporites				
Halite	$NaCl$	Cubic	58.45	2.16
Anhydrite	$CaSO_4$	Orthorhombic	136.2	2.96
Gypsum	$CaSO_4 \cdot 2H_2O$	Monoclinic	172.1	2.32

materials in chemical, glass, and metallurgical industries; and as construction materials in building stones and ingredients in plaster and concrete.

(*b*) *Shales* are defined as fine-grained rocks composed of particles with diameters smaller than 1/16 mm and form approximately 50% of the sedimentary rocks. Clay minerals (table I.3) are the dominant constituents of shales (fig. I.4), but fine-grained quartz and feldspar are also common. Despite their abundance they are not found exposed as often as the more resistant limestones and sandstones. Because of their complex composition and fine particle size, shale properties are not as well understood as those of other sedimentary materials. Despite this factor, shales, and clays in particular, have found great use in economic enterprises: as raw materials in cements, roofing tiles, bricks, pottery, and ceramic ware. Due to their extremely small particle sizes and low permeability, they are used as container linings for the long-term underground disposal of wastes.

(*c*) *Carbonate* sediments are composed primarily of the minerals calcite, aragonite, dolomite, and some magnesium carbonates (table I.3). Their origin can be organic (biogenic skeletal remains of living marine organisms) or inorganic (chemical precipitates from water). They constitute approximately 20% of all observed sedimentary rocks. Many carbonate rocks have formed as a result of the precipitation of calcite or aragonite by living organisms and the subsequent incorporation of the remaining skeletal fragments of these organisms into sedimentary deposits. An average limestone will have calcite as the dominant mineral phase, $\sim$ 80%, with an admixture of other magnesium carbonates, $\sim$ 20%. Quartz and feldspar are minor constituents and do not exceed 10% of the rock.

(*d*) In contrast to the detrital sediments sandstone, shale, and carbonate, *evaporites* are chemical sediments. Evaporites result from evaporative concentration of a water body and subsequent precipitation of supersaturated components. In many cases the evaporating medium is sea water and the chief minerals formed are gypsum, anhydrite, and halite (table I.3). Thick deposits of these minerals are common; however, they constitute only a small percentage (5%) of all sedimentary rocks. Because evaporites are chemically precipitated sediments, they are formed with continuous contact between grains and insignificant amounts of pore space. From this viewpoint their mode of formation and microstructure is closer to that of igneous rocks, which can be viewed as high-temperature precipitates from silicate melts.

Grain Size and Grain Size Distributions

Much effort has been expended on the classification of sedimentary rocks. For clastic sediments one of the most important microstructural parameters is grain size. For example, in chapter V we shall discuss fluid transport through a porous medium, which is characterized by a property called the permeability of a rock. For the permeability the grain diameter is an essential parameter, because grain diameter affects pore size. Similarly, grain size is inversely related to the grain surface area. Because reactivity of sedimentary rocks is proportional to surface area, it is also strongly influenced by the mineral grain size. Many naturally occurring sediments contain a range of particle sizes when they are initially deposited. In a subsequent section we shall discuss how a distribution of particle sizes (sorting) affects the porosity and rock microstructure. For these reasons it is useful to have a perspective on the range of particle sizes that occur in sedimentary rocks and how the grain size is used in the classification of clastic sediments.

The size distribution of sedimentary particles is quite enormous, ranging from clay particles of 0.01 μm (10^{-8} m) to conglomerates which

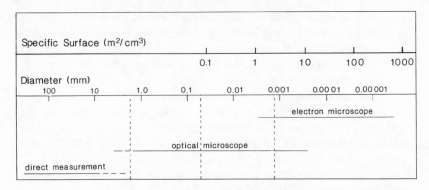

Fig. I.6. Grain diameters and specific surfaces for sedimentary rocks. (After Pettijohn, 1975.)

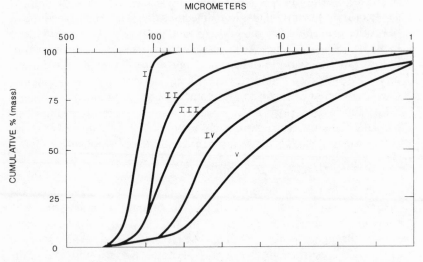

Fig. I.7. Distributions of grain diameters for recent sediments. I—Beach and shallow-water sands (coarsest sediments). II—Primarily sands with some silts and shales (deposited in zones with strong currents). III—Sands (50%) and silty-shales (50%), (deposits formed at the borders of principal currents and in moderate water depth). IV—Silts with some organic material. V—Very fine sediments with a large percentage of organic matter (deposits formed in zones far removed from the main currents). (After Krumbein, W. C., and E. Aberdeen, 1937. The Sediments of Barataria Bay, *J. Sed Pet*. 7:3–17.)

contain grains 10 cm (10^{-1}m) in diameter: a range of sizes encompassing seven orders of magnitude (fig. I.6). The specific surface area is defined as the surface area of the pores divided by volume of solids. For the simple case of an assemblage of identical spheres, the specific surface is $(\pi d^2)/(\pi d^3/6) = 6/d$ where d is the grain diameter. Figure I.6 shows the inverse relation between these two quantities. For comparison purposes note that the limit of resolution by x-ray diffraction is 0.004 μm, that by electron transmission microscope .0005 μm, while individual $CaCO_3$ molecules and SiO_2 tetrahedra are approximately 0.0003 μm.

Many factors determine the distribution of grain diameters. At the time of deposition, both hydrodynamics (the depositional process itself) and the sediment source are important. Fig. I.7 shows particle size distributions found in recent sediments. The five distributions shown reflect water energy in the depositional environment, with (I) being the highest energy and (V) the lowest. As the energy of the environment decreases, the particle size distribution is wider (sorting becomes poorer) and the distribution is shifted toward smaller particle sizes. Transport of very fine particles results from viscous drag forces which are proportional to the particle surface area (Stokes' Law). Because the ratio of viscous force to gravitational force on a particle increases as its diameter decreases, smaller particles remain in suspension longer. There is an upper and lower limit to the size of sedimentary particles. If they are too large, they cannot be easily transported, and if they are too small, they will not settle out of the transporting medium.

PROBLEMS

I.a. Rate of Dissolution and Mineral Lifetimes

The rate of dissolution, in moles $m^{-2}s^{-1}$, is a specific constant k for each mineral species. The table below shows values for the rate con-

Mineral	k (moles m^{-2} s^{-1})	v (m^3 mole^{-1})
Quartz	$4.1 \cdot 10^{-14}$	2.4×10^{-5}
Muscovite	$2.56 \cdot 10^{-13}$	4.8×10^{-5}
Forsterite	$1.2 \cdot 10^{-12}$	4.6×10^{-5}
K-Feldspar	$1.67 \cdot 10^{-12}$	3.8×10^{-5}
Albite	$1.19 \cdot 10^{-11}$	3.5×10^{-5}
Enstatite	$1.0 \cdot 10^{-10}$	3.8×10^{-5}
Anorthite	$5.6 \cdot 10^{-9}$	5.3×10^{-5}

stant k (at $T = 25°C$, pH = 5) and the molar volumes v for various silicate minerals. The number of moles dn per unit time dt going into solution is $\dfrac{dn}{dt} = kA$, where A is the surface area of a grain. Assuming the grains are spherical and of radius r, compute the lifetime of the different minerals for the conditions pH = 5, $T = 25°C$.

REFERENCES

Aubouin, J., R. Brousse, and J. P. Lehman. 1968. *Precision in Geology*. Vol. I: *Petrology*. Paris: Dunod.

Smith, D. G. 1981. *Cambridge Encyclopedia of Earth Science*. Cambridge: Cambridge University Press.

Garrels, R. M., and F. T. MacKenzie. 1971. *Evolution of Sedimentary Rocks*. New York: W. W. Norton.

Pettijohn, F. J. 1963. Chemical Composition of Sandstones—Excluding Carbonate and Volcanic Sands. *Data of Geochemistry*, 6th ed., M. Fleischer, ed. USGS Profess. Paper, 440-S.

———. 1975. *Sedimentary Rocks*. New York: Harper.

Turner, F. J., and J. Verhoogen. 1960. *Igneous and Metamorphic Petrology*. New York: McGraw-Hill.

II. Porous Media

PRACTICALLY all macroscopic physical properties of rocks are influenced by the pore microstructure. Depending on the type of porous medium, porosity can vary from close to zero in igneous crystalline rocks, to almost one for fibrous man-made materials. Ideally, if we had complete knowledge of the pore space spatial distribution, we could describe this distribution mathematically through a function $\Omega(x, y, z)$ which takes on the value $\Omega(x, y, z) = 1$, if the spatial position (x, y, z) is located in a pore, or $\Omega(x, y, z) = 0$, if (x, y, z) is in the solid. This pore distribution function Ω contains all information about the pore structure of a rock. For example, the pore volume of the rock would simply be:

$$V_p = \int \Omega(x, y, z)\, dV$$

where the integral is over the whole rock volume.

Clearly, such detailed microscopic information is normally not available or measurable. Therefore, there has been an emphasis on certain measurable macroscopic parameters which convey much of the essential information about pore structure. These macroscopic parameters are the porosity, specific surface area, imbibition and drainage curves, electrical formation factor, and permeability. Porosity is defined as the pore volume per unit rock volume and is proportional to the average of the pore space distribution function Ω. The specific surface area of the pore space contains information on the pore sizes, that is, the short-range correlations in the pore structure. Capillary pressure function, electrical formation factor, and permeability all contain some information on the long-range correlations in the pore structure such as the size distribution and connectivity of the pores. Because the electrical formation factor and permeability can be viewed as derived quantities rather than direct measurements of pore structure, they will be discussed in detail in subsequent chapters which address the individual physical properties of rocks. In this chapter we shall focus on the primary macroscopic parameters describing pore structure: porosity, specific surface area, and the imbibition-drainage curves.

POROSITY

Few properties of natural materials have received more attention than porosity and permeability. Porosity is a measure of the pore volume available within the rock, and permeability indicates the ease with which fluids can flow through this pore space. Porosity ϕ is defined as the fraction of rock volume V that is not occupied by solid matter. If the volume of solids is denoted by V_s and the pore volume as $V_p = V - V_s$, we can define porosity as

$$\phi = \frac{V - V_s}{V} = \frac{V_p}{V} = \frac{\text{Pore Volume}}{\text{Total Volume}}.$$ II.1

Note that porosity does not give any information concerning pore sizes, their distribution, and the degree of connectivity. Thus rocks which have identical porosities can have widely different physical properties, such as permeability. In most problems involving porous rocks or soils, it is useful to reference all volumes to the volume of solids V_s. Thus the pore volume per unit volume of solids, V_p/V_s, called the *void ratio*, is

$$\frac{V_p}{V_s} = \frac{\phi}{1 - \phi} = \frac{\text{Pore Volume}}{\text{Solid Volume}}.$$ II.2

One major difference between igneous and sedimentary rocks is seen in their porosity structure. This difference arises due to the distinct processes of formation of these two rock groups.

Igneous Rocks

At the time of formation, intrusive rocks do not have any significant intergranular porosity. Formed by crystallization from liquid magma, the grains intergrow tightly, leaving almost no void space. Typically, granite after formation has minimal porosity, $\phi \sim 10^{-3}$, most of which occurs as small irregular cavities which are remnants of the crystallization process. Volcanic rocks are quite different. Due to the rapid cooling at the surface and exposure to weathering elements, their porosity is greater and exhibits more variability. Fluid transport through igneous rocks occurs mainly through cracks and fractures which have arisen later in the rock's history in response to tectonic or thermal stresses. The degree of fracturing is highly variable and can be observed at all length scales. Fractures and fracture spacings can be seen on the microscale, on the scale of a few centimeters, and up to tens of meters. In many cases fracture distributions contain fractures of all sizes, and the fracture pattern "appears" the same at all scales of observation. These types of distributions which appear invariant on a change of scale are called "fractal" (Mandelbrot, 1983). Fractures are "planar" discontinuities which traverse the rock mass but occupy a very small fraction of the rock volume. In most cases fractures intersect each other forming an interconnected network through the rock. In a highly fractured igneous rock the porosity can be quite low, on the order of 10^{-2}, yet this small fracture porosity can have a very high permeability.

Sedimentary Rocks

The mode of formation of sedimentary rocks (accumulation of debris or chemical precipitation) explains why certain rocks occur with no significant porosity and others with porosities greater than 0.50. Clastic sedimentary rocks are composed of transported and mechanically deposited individual grains and thus are formed with a continuous three-dimensional porosity network. The initial porosity of a clastic sediment depends somewhat on particle size, but much more so on the distribution of particle sizes. Sediments with a narrow distribution of particle sizes, such as well-sorted river bed sands, have initial porosities in the range 0.40–0.45. For comparison, the porosity of a random packing of uniform-sized spheres is 0.40. The dependence of initial porosity on grain diameter is due to the ratio of gravitational to frictional forces acting on the grains. As the grain size decreases, friction forces become comparable to the gravitational force acting on a particle at an earlier stage of settlement. Thus smaller-size particles form a rigid sedimentary framework with a larger initial porosity. When a sediment is buried, the increasing stress compacts the sediment to smaller volumes, with the volume change being primarily a loss of porosity. Compaction is a result of the stress concentrations at grain contacts and the subsequent grain deformation by the displacement of mass from contact points to free surfaces through a dissolution-precipitation process.

Chemical sediments that form due to the evaporation of seawater (evaporites) generally have extremely low porosities (10^{-3}). This again is a consequence of their mode of formation: crystal growth from solution. In addition the mineral grains of evaporites, such as halite, deform readily under non-hydrostatic stress conditions, causing the sediment to compact and eliminate most of the pore space early in its history. We will return to the processes of deformation and compaction in later chapters.

Methods of Measurement

Methods for measuring porosity have now been standardized and have been reviewed in detail in many textbooks (i.e., Scheidegger, 1974). Therefore, only a brief outline of the various methods will be presented.

Direct method: The two volumes, V and V_s, that appear in equation II.1 are determined directly. Then $\phi = 1 - (V_s/V)$ gives the average porosity of the rock. Because the direct method determines the pore volume from the difference $V - V_s$, it will include all of the pore space even when it is not connected to the outside of the rock by a continuous pore network.

Imbibition method: Immersing the porous sample into a perfectly wetting fluid for a sufficiently long time will cause the fluid to enter (imbibe) into all of the pore space that is connected to the rock exterior. If the sample is weighed before and after imbibition, the difference in weight will be ρV_p, where ρ is the known fluid density and V_p is the pore volume. A volumetric displacement measurement on the saturated sample will give the bulk volume V of the sample. Then $\phi = V_p/V$. Imbibition will yield the best values for the connected porosity.

Mercury injection: After evacuating a rock of pore fluids, the bulk volume is determined by immersing the rock in a mercury bath. For most earth materials, mercury will not enter the pore space at atmospheric pressure conditions. The mercury pressure is then raised until it begins to enter the pore space. If the pressure is raised sufficiently, mercury will enter most pores, thus determining the pore volume. Usually there will always be an "irreducible" amount of pore volume that is not accessible to mercury, because it would require extreme pressures for penetration. The relation between the volume of mercury that has drained into the pore space and applied pressure will be discussed later in this chapter.

Gas expansion: Conceptually this method is similar to mercury injection except that the medium used is highly compressible (ideal gas) and requires almost no pressure differential for entering the pore space. First the bulk volume V of the rock is determined through an independent measurement. Then the rock is enclosed in a container of known volume V_1, having initial gas pressure P_1, and connected to an evacuated container of known volume V_2. When the valve between these two containers is opened slowly, isothermally, the gas expands into the evacuated container and the gas pressure decreases to P_2. The pore volume V_p can then be determined using the ideal gas law: $P_1(V_1 - V + V_p) = P_2(V_1 - V + V_p + V_2)$.

Density methods: Measuring the bulk density ρ of the rock and the average density ρ_s of the solid constituents allows the determination of the total porosity $\phi = 1 - (\rho/\rho_s)$. For example, if the rock is formed purely of quartz grains, $\rho_s = 2.65$ gm/cm^3, then measuring the bulk rock density ρ will yield the porosity.

Optical methods: For cases where the porosity microstructure is isotropic, porosity can be determined from a two-dimensional section as $\phi = A_p/A$, where A is the area of the total image and A_p is the area intersected by the pores. Usually it is necessary to impregnate the pores with a material such as Wood's metal or a plastic to enhance the contrast between minerals and pores. A thin section can then be digitized and the porosity computed. This approach can be generalized to show the distribution of other components within a rock. If the porous media is composed of pores of two widely distinct sizes, then due to capillary forces it is possible to saturate the smaller/larger pores with two fluids of different reflectivities, one occupying the larger and the other the smaller pores. This technique could be used to show the amount and distribution of clay minerals within a porous rock composed of large quartz grains.

The preceding methods are all laboratory measurements. Well logging utilizes other techniques which permit an estimation of porosity indirectly in situ. These techniques depend on instruments being lowered into a well and the rock response to an impulse recorded. The impulse can be acoustical, electromagnetic, neutron, etc. In most cases a significant theoretical development is required for interpreting the measured rock response. This theoretical understanding is one of the principal goals of this book.

POROSITY EVOLUTION

A sediment is formed when detrital particles which have been in suspension in a fluid settle downward under the force of gravity. A suspension of sedimentary particles in a fluid can be viewed as a porous medium with a fractional porosity close to one. As gravitational settling of particles takes place, porosity decreases and the number of interparticle contacts increases. Eventually, the gravitational force acting on a particle is balanced by interparticle friction and/or cohesive forces, and the sediment is stabilized. At this point the sediment has made a transition from the liquid state with no shear strength, to a state with a measurable shear strength. A characteristic feature in the formation of a clastic sediment is the appearance of a grain framework structure which resists the gravitational force and prevents further settling. Any gravitational energy gained in further settling is insufficient to overcome the frictional/cohesional forces. Within this framework, grain-to-grain contacts support all of the load and yet involve only a small part of the grain surface area. Because of this stress-supporting grain framework, there can exist a large void volume between the grains. This is the initial porosity of a sediment. When it is later buried, the increasing pressure and temperature activate various chemical, physical, and biological processes that alter the sediment, lowering the porosity from its initial value. These diagenetic processes include the consolidation and compaction of a rock, dissolution and precipitation of minerals, recrystallization, and others. To arrive at a better understanding of the evolution of porosity in sedimentary rocks, we shall first look at the factors that control the initial porosity of clastic sediments, and later, consider the porosity changes due to burial.

Initial Porosity

The microstructural parameters affecting initial porosity are grain size, grain shape, and the distribution in grain sizes.

Grain Size: For a sediment composed of a single grain size, the force required to overcome frictional and cohesive forces increases with the exposed surface area of the grains. Because the specific surface area (exposed surface area per unit solid volume) is inversely proportional to particle size, a unit weight of fine particles will be stabilized at a larger porosity than a unit weight of coarse particles, all other factors being equal. Observations shown in figure II.1 support this general relationship between particle size and porosity. There is a trend to

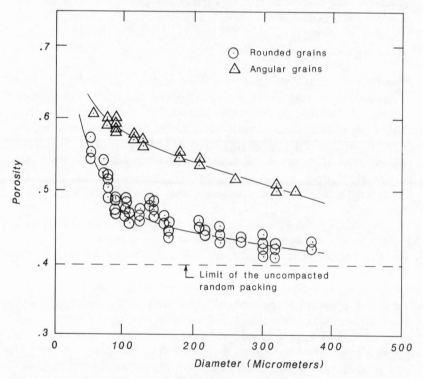

Fig. II.1. Relation between porosity and particle diameter for assemblages of uniform size grains. (After Leva, M., M. Weintraub, M. Grummer, M. Pollchik, and H. H. Storch, 1951, Fluid flow through packed and fluidized systems, U.S. Bureau of Mines Bull. No. 504.)

Table II.1. THE EFFECT OF PARTICLE SHAPE ON THE POROSITY, d = DIAMETER, l = LENGTH

Shape	Porosity
Sphere	0.399
Cube	0.425
Cylinder ($d = 1$)	0.429
Disk ($d = 21$)	0.453

Source: After Wyllie, M.R.J. and A. R. Gregory, 1955, Fluid flow through unconsolidated porous aggregates, *Industrial and Engineering Chem.* 47:1379–88.

increasing porosity as the particle size decreases, but the increase only becomes significant for particle diameters below 100 μm. For extremely fine-grained sediments (recent shales), very high porosities, in some cases approaching 0.80, have been measured. As the particle diameters become larger, the effects of friction/cohesion decrease and a limiting value for the initial porosity is reached. This limit of frictionless particle packing has been studied quite intensively, with the random packing of uniform-size spheres receiving the most attention. This resulting limiting porosity of .399 is independent of sphere diameter and appears to be a fundamental constant of statistical geometry. At this point no further loss of porosity can occur without irreversible grain deformation through processes such as dissolution-recrystallization, fracture, or plastic flow. A decrease in porosity due to grain deformational processes is called compaction.

Particle shape: The role of particle shape on porosity is not well understood. There have been several studies of packing of identical particles that are not spherical in shape. All of the nonspherical shapes that have been studied were found to have porosities larger than that for spheres. The results shown in Table II.1 are independent of particle size. They correspond to the limit where frictional and cohesive forces between particles are negligible.

Distribution of grain sizes: Sediments in less energetic environments are deposited with a wide distribution of grain sizes, for example, curve V of figure I.7. In fact bimodal distributions, distributions with maxima at two distinct diameters, occur quite frequently. The bimodal distribution contains most of the features required for understanding the consequences of particle size distributions. For this reason we shall examine in detail how the porosity of a bimodal mixture varies as the volume fraction of larger and smaller particles is changed.

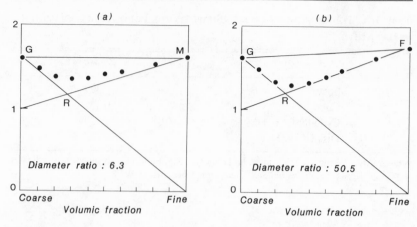

Fig. II.2. Apparent volume for assemblages of spheres of diameter ratios (*a*) 6.3 and (*b*) 50.5. (After Westman A.E.R. and H. R. Hugill, 1930, *J. Amer. Ceram. Soc.* 13:767.)

Random packings of variable diameter spheres have been studied experimentally and theoretically through computer simulation (fig II.2). Westman and Hugill (1930) measured the "apparent volume" V_a of mixtures of "round" sand particles of two distinct sizes: $V_a = V/V_s = 1/(1 - \phi)$. Results obtained in several of their experiments for two binary systems having diameter ratios 6.3 (coarse/medium) and 50.5 (coarse/fine) are shown as data points in figure II.2. Westman and Hugill noted that the sand grains were approximately spherical, but that the fine-grained sand was more angular than the coarse or medium sand. This fact would account for the somewhat larger apparent volume of the 100% fine sand and is consistent with the earlier discussion of the role of grain size and shape on initial porosity. Both cases shown in figure II.2 exhibit a decrease in V_a and ϕ as the two sizes are mixed homogeneously. This porosity decrease can be interpreted in the following manner. If the diameter ratio is unity, there would be no particle size effect on mixing and the apparent volume would follow the straight lines *GM* and *GF*. If the diameter ratio is very large, then it is possible for the small grains to be placed in the interstices of the large grain framework without any change in volume, although the volume fraction of fine particles is increasing. The lines labeled by *GR* show the trend for this process. To quantify this mixing process let the volume of coarse sand be denoted by V_1, that of the fine or medium sand by V_2, and the sum by $V_s = V_1 + V_2$. Let ϕ_1 and ϕ_2 represent the porosity of the 100% coarse-grained and 100% fine-grained packs. In the limit where

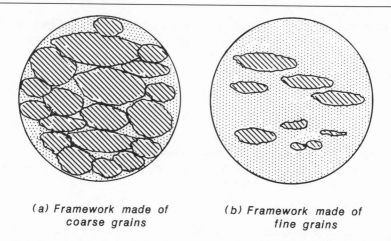

(a) Framework made of
coarse grains

(b) Framework made of
fine grains

Fig. II.3. Two possible configurations for a bimodal distribution: (*a*) framework of larger grains with the intergranular space filled with the finer grains; (*b*) larger grains "floating" in a fine-grained matrix.

the fine particles all reside in the pore space of the large-grain framework (fig. II.3a), we have $V = V_1/(1 - \phi_1)$, and $V_a = V_1/V_s(1 - \phi_1)$ is the equation defining GR. Consider now the other extreme where the grains are 100% fine (fig II.3b), and a few large grains are added. In this case a porous medium of porosity ϕ_2 is being replaced by an equal volume of solids, implying that the porosity and apparent volume will decrease. Again this can be quantified by noting that the total volume of the mixture is a sum of the volume occupied by the fine-grained porous media, $V_2/(1 - \phi_2)$, plus that occupied by the large grains V_1: thus $V = V_1 + V_2/(1 - \phi_2)$ and $V_a = \dfrac{V}{V_s} = \dfrac{V_2}{V_s}\left(\dfrac{\phi_2}{1 - \phi_2}\right) + 1$. The apparent volume decreases as the volume fraction of fine particles decreases, with the slope of the lines RM and RF now determined by the void ratio of the fine sand, $\phi_2/(1 - \phi_2)$. The observed data points lie within the triangles GFR and GMR which form the theoretical bounds. Note also that as the diameter ratio increases from 6.3 to 50.5 the data points approach closer to the lines GRM and GRF. V_a has a minimum at R which occurs at approximately 20 to 30% of the smaller particles. The porosity at R can be estimated by assuming that the porosity of the large-grained sand, 0.40, is totally filled by a fine-grained porous sand of equal porosity. Thus the porosity at R is $(0.4)^2 = 0.16$.

Sandstones found in nature have a great variety of grain shapes and size distributions, and thus initial porosities. Beard and Weyl (1973) made a comprehensive study of the relation between porosity and grain size distributions. They found that mean porosities ranged from 0.279 (very poorly sorted) to 0.424 (extremely well sorted), with the porosity increasing monotonically with increased sorting. From these experiments the following important conclusion can be made: the maximum initial porosity will always occur for "perfectly" sorted sediments (i.e., uniform grain size) and the porosity will depend on the shape of particles and grain size; but it will not be lower than 0.40, the value for a random packing of spheres. These concepts of particle packing and particle size distributions have been applied to interpreting variations in reservoir properties (Clarke, 1979), but are applicable to any mixture with a particle size distribution.

Effects of Burial: Compaction

As a sediment is buried, the overlying sediments increase the effective stress on the grain framework causing particle movement and grain deformation to take place. Effective stress is a fundamental concept of soil mechanics introduced by Terzaghi (1936). For unconsolidated soils it is approximately equal to the differential stress, that is, the difference between the total confining stress σ_{ij} and the pore fluid pressure P (see chapter V).

$$\text{Differential Stress } \sigma_{ij}' \equiv \sigma_{ij} - P\delta_{ij}. \qquad \text{II.3}$$

Here $\delta_{ij} = 1$, if $i = j$, and zero otherwise. Differential stress is a measure of the average stress above the fluid pressure that is supported by the grain framework. It is the driving force for rock deformation. If the total confining pressure is equal to the fluid pressure, $\sigma_{ij} = P\delta_{ij}$, the differential stress is zero, and the mineral grains are then only subjected to a uniform confining pressure equal to the fluid pressure.

As the differential stress increases with burial, it overcomes the frictional/cohesional forces holding the framework together and particle sliding and rotations take place. This is accompanied by pore fluid expulsion and a decrease in rock volume. These processes of mechanical rearrangement or *consolidation* do not involve irreversible grain deformation. Consolidation continues until no further volume decrease can be achieved without some permanent grain deformation. This limiting volume is often referred to as the Atterberg plastic limit (ASTM, 1964). For clays this limit is reached when the compressive effective stress is in the range of 20 to 60 bars which is normally attained at subsurface depths in the range of 250 to 800 meters. For coarse-grained sediments such as sands, frictional forces are not as important and little consolidation takes place. Most of the volume decrease in sandstones occurs through irreversible grain deformation, which is referred to as *compaction*. Porosity evolution in a sedimentary rock will depend on the rate of stress increase with time, the particular microscopic process that is causing grain deformation, and the rate at which the pore fluid can be expelled.

Although the microscopic processes involved are quite complex and the geologic history varies from location to location, there are several generalizations that can be made. On the average, the initial porosities of shales and other fine-grained sediments are higher than those of sands, but their consolidation and compaction is more rapid so that at depths between 3–5 km, they have porosities somewhat less than those of sands. These observations are based on average measured porosities. But, it should not be assumed that two sandstones or two shales that are at or have been at the same maximum depth will have the same porosity. Two sandstones at the same depth z can have different porosities, due to varying amounts of fine-grained minerals present within the intergranular pore volume. These minerals could have been present at deposition or could have been precipitated later as cements. If porosity only depended on sorting and the maximum effective stress acting on the grains, a sand composed of particles of uniform size would have the maximum porosity for given stress conditions. This maximum porosity varies with depth z (fig. II.4), where the function $\phi(z)$ was obtained by assuming that the grains can only sustain a maximum critical effective stress ϕ_c at contacts before rapid deformation occurs (Palciauskas and Domenico, 1989). On the average, stress increases very slowly during burial, on the order of 13 bars/million years. Grain deformation on the other hand is relatively rapid, so that the grain contact stress is maintained at its critical value σ_c. Any increase in the sediment load thus results in an increase in grain contact area and a decrease in volume. This critical stress depends on the latent heat of fusion h_m and molar volume v_m: $\sigma_c \cong h_m/v_m$. Figure II.4 also shows porosity-depth data for 17,367 sandstone cores from southern Louisiana. Because these sandstone cores contain undetermined amounts of pore-filling clays and cements, the measured porosities should be lower than the curve for the maximum, but ideally should have a similar slope with depth. This data is typical of sandstones.

Other Diagenetic Processes

Diagenesis encompasses all of the chemical, physical, and biological processes that alter a sediment from the moment of formation. We have briefly discussed one of the major diagenetic processes, compaction, but several other processes can be of equal importance. The topic of diagenesis is much too broad to be covered in this book. Therefore, only several important physical-chemical processes will be discussed briefly.

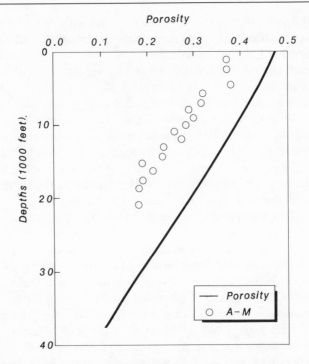

Fig. II.4. Theoretical upper limit to the porosity $\phi(z)$ of sandstone. The 17 experimental points have been obtained by averaging 17,367 core measurements of tertiary sandstones. (From an unpublished report by G. I. Atwater and E. E. Miller, 1965, presented by H. Blatt, 1979, Diagenetic Processes in Sandstones, in P. A. Scholle and P. R. Schluger, eds., Aspects of Diagenesis: SEPM, Special Publication 26:141–57.)

A major process is rock-fluid interaction. If the moving pore fluid is not in chemical equilibrium with the rock-forming minerals, dissolution/precipitation reactions will take place. This can result in long-range mass transport of certain species before they are precipitated as cements on the surface of grains. Carbonate and silica are the major cements in sandstones, with clays such as kaolinite and chlorite also being common. Cementation decreases the porosity and permeability. On the other hand, surface area increases due to the morphology of mineral growth. In some cases the framework grains are dissolved and recrystallized locally on the surfaces of other framework grains without a change in mineralogy (cf. chap. VI). This decrease in the grain framework volume implies a local transfer of mass from the stress-bearing framework to cement, implying compaction and a decrease in porosity. In other cases a different mineral is locally precipitated from the one dissolved (such as dolomite from calcite, when an ample amount of magnesium is present). The difference in volume between the initial calcite grains and the recrystallized dolomite accounts for the increase in microporosity (the molar volume of dolomite is 12.3% smaller than calcite). All of these diagenetic reactions alter the microstructure and physical properties of a rock.

SURFACE ENERGY

In our daily experiences we continuously observe well-defined surfaces which separate various phases of matter: liquid-liquid, liquid-gas, solid-gas, solid-liquid, and solid-solid. We are also familiar with many capillary phenomena associated with surfaces, such as the rise of a liquid in a cube of sugar when the lower end of the sugar cube is dipped into the free surface of the liquid. These types of phenomena clearly show that there are forces other than gravity acting on the liquid: these are the forces due to the presence of a surface (surface tension). Surface effects play a significant role in the thermodynamics of a system when the free energy of a surface becomes comparable to the volume free energy. For a spherical droplet of fluid, the ratio of surface area/volume is inversely proportional to the radius. Thus, when a droplet becomes "sufficiently" small, its surface energy cannot be neglected and plays an important role in the equilibrium properties and dynamics of a fluid. This is just the situation in porous rocks where the pores are of microscopic size and are often saturated with two immiscible fluids. In these cases the physical properties of a rock can also be strongly modified.

Specific Surface Area

The "specific surface" area of a porous media is defined as the surface area of the pores per unit volume of solids. Specific surface area plays an important role in the medium's adsorption capacity, catalysis, and ion exchange. In certain cases the specific surface can be related to permeability. All of these observations result from the fact that the specific surface area is inversely proportional to capillary size. For the simple case of equal-sized spheres of diameter d, $S = \dfrac{6}{d}$. For a network of cylindrical capillaries all of radius r, $S = \dfrac{2\phi}{r(1 - \phi)}$. When there is a mixture of grain (capillary) sizes, one being much larger than the other such as clays in a sand, the specific surface area is primarily determined by the surface area of the finer particles. The major methods for determining specific surface have been discussed by Scheidegger (1974).

Surface Tension

In our daily experiences we often see that an interface between two immiscible homogeneous fluids responds like an elastic membrane. Perhaps the most common experience is the observation that soap bubbles, after formation, tend to contract in size as if their surface is in a state of tension, and that this contraction continues until the surface is stabilized by the internal pressure at the smallest possible surface area. Forces causing this contraction arise from molecular interactions at the microscopic level. A molecule in the interior of a fluid experiences attractive interactions with the surrounding molecules, thereby lowering its energy. The molecules on the surface only experience attractive interactions from the interior molecules, resulting in a higher molecular energy on the surface and a force of attraction toward the interior of the fluid. Any attempt at extending or enlarging the surface is resisted by this force of attraction. Equivalently, the higher energy in the surface layers implies that work dW must be done on the system to increase the surface area by dA. From a thermodynamic point of view, under isothermal conditions

$$dW = \gamma \, dA. \qquad\qquad\qquad \text{II.4}$$

The coefficient γ is called the surface energy (joules/m^2 or ergs/cm^2) or surface tension (Newtons/m or dynes/cm). The relationship between units is as follows:

$$1 \text{ Joule/m}^2 = 1 \text{ Newton/m} = 10^3 \text{ ergs/cm}^2 = 10^3 \text{ dynes/cm}.$$

Normally the surface tension is positive, $\gamma > 0$, and it takes a positive amount of energy to form a surface. If the surface energy is negative, $\gamma < 0$, then it is energetically favorable to increase the surface area without limit. The two phases would mix, continually increasing their interfacial area: the two phases are miscible. The mathematical equivalence of surface energy and a tension acting in the surface may be illustrated by a film on a wire frame (fig. II.5), where the arm AB can move without friction. To prevent the film from contracting, a force G must be applied parallel to the film surface and normal to AB as shown. If the arm is displaced very slowly through a distance dx, the surface area increases by $2h \, dx$ (the film has two surfaces). The work done is $dW = G \, dx$. Similarly, the work required to increase the surface area by $2h \, dx$ is $\gamma 2h \, dx$. Equating these two expressions, $G \, dx = \gamma 2h \, dx$, allows the interpretation of γ as, $\gamma = \dfrac{G}{2h}$, the mechanical force per unit length (surface tension).

Let us examine the case of a curved surface separating two fluid phases which are at pressures P_1 and P_2, respectively. A simple

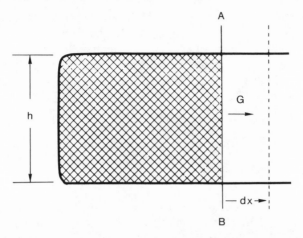

Fig. II.5. Equivalence of surface energy and surface tension.

example is a spherical inclusion of radius r of fluid (1) in fluid (2). Assume that the two fluids occupy a total volume, $V = V_1 + V_2$, and are at equal temperature $T = T_1 = T_2$. At equilibrium, the thermodynamic potential of the combined system, two phases plus interface, is at a minimum, and the work done by an infinitesimal change in the radius of the sphere, dr, is equal to zero.

$$-P_1 \, dV_1 - P_2 \, dV_2 + \gamma \, dA = 0.$$

The inclusion occupies a volume $V_1 = (4/3)\pi r^3$ and has a surface area $A = 4\pi r^2$: thus $dV_1 = -dV_2 = 4\pi r^2 \, dr$ and $dA = 8\pi r \, dr$. The preceding equation becomes:

$$P_1 - P_2 = \frac{2\gamma}{r}. \qquad \text{II.5}$$

This is the first fundamental equation of capillarity: Laplace's equation. It states that due to the existence of a surface tension, two fluids at equilibrium can be maintained at different pressures. Note that the interior pressure is larger, and that if the interface is planar ($r \to \infty$), the two pressures become equal. Equation II.5 was derived for the case of a spherical surface, but it can be generalized to any point of a nonspherical surface by replacing r with the mean radius of curvature r_m:

$$\frac{1}{r_m} = \frac{1}{2}\left(\frac{1}{r_1} + \frac{1}{r_2}\right)$$

where r_1 and r_2 are the two principal radii of curvature at the point considered (Defay and Prigogine, 1966). The previous discussion assumes that gravitational effects are negligible. In this case the two pressures are uniform throughout the two phases and the mean curvature r_m of the interface is constant at every point of the surface. When gravity is not negligible, the pressures P_1 and P_2 will vary with height and the radius of curvature will not be constant. Equation II.5 is a force balance relation involving P_1, P_2, γ, and r_m, and is applicable at *all* points along the interface even when these parameters vary along the interface.

Wettability

Because the surface of a solid does not have the mobility of a fluid interface, it is difficult to make a direct measurement of solid-fluid surface tension. Nevertheless, there exists a surface tension between a fluid and solid, just as there exists a surface tension between immiscible fluids. When two fluids are in contact with a solid surface, the equilibrium configuration of the two fluid phases will depend on the relative values of surface tension between the three phases (fig. 11.6). Let γ_{sl}, γ_{lg}, and γ_{sg} denote the surface tension between solid and liquid, liquid and gas, and solid and gas. Each of the three surface tension forces acts in the interface between the corresponding phases, and all three surface tension forces act perpendicular to the line where the three bodies meet. θ is defined as the angle of contact between the liquid surface and the plane surface of the solid. Mechanical equilibrium implies that the sum of the projections of the surface tension forces parallel to the solid surface must sum to zero.

$$\gamma_{lg} \cos \theta = \gamma_{sg} - \gamma_{sl}. \qquad \text{II.6}$$

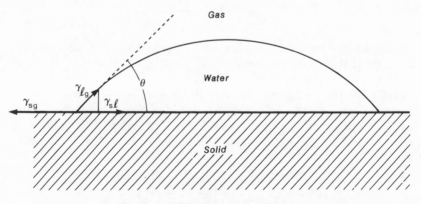

Fig. II.6. Solid-liquid angle of contact.

Table II.2. CONTACT ANGLES θ AND INTERFACIAL TENSION γ FOR SOME COMMON FLUID-FLUID INTERFACES

Interface	θ (degrees)	$Cos\theta$	γ (dynes / cm)
Air-Water	0	1.0	72
Oil-Water	30	0.866	48
Air-Oil	0	1.0	24
Air-Mercury	140	-0.765	480

This equation, established qualitatively by Young in 1805, is referred to as Young's equation. It was presented in algebraic form by Dupre in 1869 and is the second fundamental equation of capillarity. Note that if $\gamma_{sg} > \gamma_{sl}$, then $\cos \theta > 0$ and $\theta < 90^0$; while if $\gamma_{sg} < \gamma_{sl}$, then $\theta > 90^0$. For a stable contact we must have $|\cos \theta| \le 1$ or equivalently $|\gamma_{sg} - \gamma_{sl}| \le \gamma_{gl}$. This inequality is not satisfied when $\gamma_{lg} + \gamma_{sl} < \gamma_{sg}$, when the liquid covers the whole area of the solid. Alternatively, we can have $\gamma_{lg} + \gamma_{sg} < \gamma_{sl}$, in which case the solid-liquid contact is completely displaced by the gas-solid surface (see table II.2). For a more comprehensive discussion of contact angles, their measurement, and the interpretation of surface tension of a solid-fluid interface see the text of Adamson (1982).

When one fluid tends to spread along the solid interface and exclude the second fluid, the first is called the "wetting" fluid and the other "nonwetting." On a microscopic scale, wettability arises from the relative strength of molecular interactions between each fluid with the solid. The wetting fluid has stronger interactions with the solid, and thus tends to spread, minimizing the total energy of the system. The macroscopic manifestations of these molecular interactions can be quite varied. Figure II.7 shows four possible configurations of oil and water in contact with a solid surface. Figure II.7c shows an angle of contact less than 90 degrees and is called water-wet, while figure II.7d shows the extreme case of total wettability, $\theta = 0$. Thus a statement that a surface is water-wet is only a statement of preference of contact and does not necessarily imply that water is everywhere in contact with the solid.

IMBIBITION AND DRAINAGE

Capillary Pressure

Capillary pressure is the pressure difference across an interface separating two immiscible fluids, in the interior of a capillary. It is a parameter which depends on the capillary structure, and in the general case of a

$$\gamma_{\ell g} \cos \theta = \gamma_{sg} - \gamma_{s\ell}$$

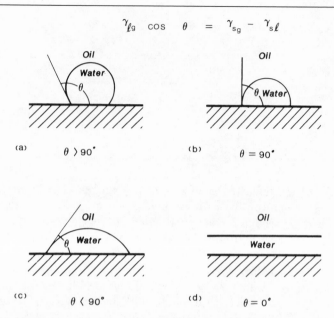

Fig. II.7. Wettability: (*a*) the wetting liquid is oil; (*b*) neutral wettability; (*c*) the wetting liquid is water; (*d*) total water wettability.

porous rock, on the pore microstructure. As an illustration, consider the simple example of how high a liquid can rise in an open capillary of radius R, when the lower end of the capillary is placed in a beaker filled with fluid (fig. II.8). If the liquid is the wetting fluid, the liquid pressure will be less than gas pressure and the interface will be concave to the fluid. Neglecting the effects of gravity on the interface, the radius of curvature r_m is a constant and equal to $R/\cos \theta$. Laplace's equation II.5 gives the capillary pressure P_c:

$$P_c = P_o - P_l = \frac{2\gamma}{r_m} = \frac{2\gamma \cos \theta}{R} \qquad \text{II.7}$$

where P_o is the gas (air) pressure above the meniscus and P_l the pressure in the fluid (water) just below the meniscus. The column of fluid of height h is supported by the pressure difference $P_o - P_l = \Delta \rho g h$ where $\Delta \rho$ is the density difference between water and air. This example illustrates the macroscopic consequences of intermolecular interactions between the solid, liquid, and air in the immediate vicinity of the contact line. The intermolecular interactions determine the contact angle θ, which in turn affects the mean radius of curvature $r_m = R/\cos \theta$, which influences the capillary pressure $P_o - P_l$, and thus the elevation h.

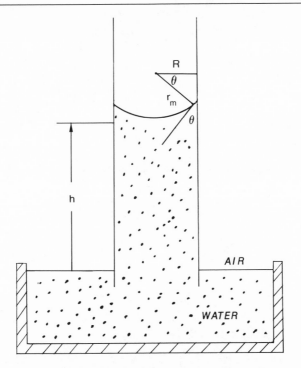

Fig. II.8. Capillary ascension by a wetting liquid.

There are many difficulties with the measurement of contact angles. Several of the methods used and their interpretations are discussed by Adamson (1982). One serious difficulty is that the contact angle in oil-water-solid systems can change with time. Craig (1971) observed that the contact angle increased with time and only after a hundred hours or so approached an equilibrium value. This dramatic change from water-wet to oil-wet conditions is a striking example of precautions that must be taken in gathering and interpreting this type of data.

Imbibition and Drainage

The preference of a solid for one fluid can cause the displacement of a nonwetting fluid by a wetting fluid without the application of an externally applied pressure. This process is called capillary *imbibition*. Consider a water-wet capillary of a very small diameter R which is initially filled with air. When the tube is brought into contact with a reservoir of water at atmospheric pressure, the initially vertical interface will be deformed by surface tension forces at the contact with the solid walls. These forces deform the interface bringing it in line with the equilibrium wetting angle θ, thereby balancing the capillary pressure and

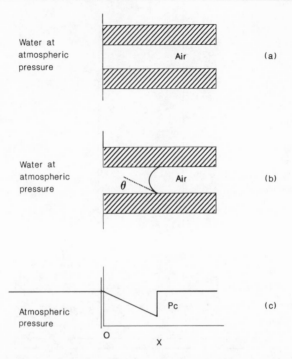

Fig. II.9. Capillary imbibition. (After Morel-Seytoux, H. J., 1969, Flow of Immiscible Liquids in Porous Media, in *Flow Through Porous Media*, ed. by R.J.M. De Wiest, N.Y.: Academic Press.

surface tension forces (fig. II.9). The pressure in the air can be taken to be uniform and atmospheric. Laplace's equation implies that the fluid pressure just behind the interface will be lower than atmospheric by the capillary pressure $P_c = 2\gamma \cos \theta/R$. There will be a pressure drop P_c between the entry point ($x = 0$) and the interface ($x > 0$), which will drive the water into the capillary. As the meniscus penetrates further into the capillary, the pressure gradient, $\dfrac{dP}{dx} \approx \dfrac{P_c}{x}$, decreases with increasing x and the flow rate will decrease with time. The average velocity of the fluid in the capillary, $\bar{v} \equiv \dfrac{dx}{dt}$, is given by Poiseuille's law

$$\bar{v} = \frac{dx}{dt} = \frac{dP}{dx}\frac{R^2}{8\mu}$$

where μ is the viscosity of the fluid. Expressing the pressure gradient through Laplace's equation yields the result

$$\bar{v} = \frac{R\gamma \cos \theta}{4x\mu}.$$ II.8

Because the average velocity is proportional to the radius, larger capillaries will be filled more rapidly than the smaller ones. Integrating equation II.8 yields

$$x^2 = \frac{R\gamma t \cos \theta}{2\mu}.$$ II.9

The position of the meniscus varies as $\sim (Rt)^{1/2}$. If instead of a single capillary we had considered a group of parallel capillaries all of the same length but with a distribution of radii, then the water would advance unevenly, saturating the capillaries in the order of the size of their radii. The progression of water is not due to an external pressure gradient but on preferential wettability.

Although this example of a single (or a group) of capillaries of uniform cross-section is very appealing due to its simplicity, it is not capable of describing many of the observed phenomena of fluid flow in rocks. Pore and pore throat sizes are in general highly variable. In addition, the pore space is multiply connected: there are many possible pathways connecting any two pores in a rock. This topological characteristic of the pore space has important consequences. During the immiscible displacement of one fluid by another, the invading fluid can bypass the initial saturating fluid, leaving it behind in the pore space. This effect can be illustrated by considering two parallel capillary tubes of different radii branching from the same beginning and later coming together (fig. II.10). Because the water enters the larger capillary more rapidly, it will reach the connection at x_1 through the larger tube first. At this point the water would enter into the smaller tube, thus trapping the gas. Although this example illustrates the basic idea of a multiply connected pore network leading to a "residual saturation," it does not consider the stability of the trapped fluid.

A clastic sediment is a porous medium composed of a network of pores, which are connected by pore throats of various sizes. It is not possible to describe the time evolution of water-air or water-oil interfaces in every capillary, without the aid of computer simulations. Capillary pressure is intimately related to the presence of a solid surface and, in particular, to the shape and size of the capillary. Normally it is not appropriate to speak of a capillary pressure of a porous media, but only of the pressure difference across a particular interface within a capillary. At equilibrium however, when the effects of gravity can be neglected, the pressure is uniform in both fluid phases, and then it is

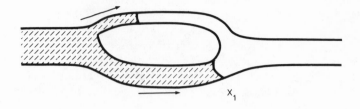

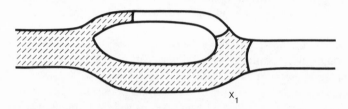

Fig. II.10. Trapping of fluid 2 during imbibition of fluid 1.

reasonable to speak about a single capillary pressure. As a specific example consider a water-wet rock, fully saturated with water, placed in contact with an oil reservoir. Because the rock is preferentially water-wet, capillary forces prevent the oil from entering the rock spontaneously. But, as the oil pressure is slowly raised by a small increment ΔP, a volume of oil will enter all accessible pores which are connected to the outside oil reservoir by apertures of radius not smaller than $R^{(1)}$. $R^{(1)}$ is related to the pressure increment ΔP through Laplace's equation: $\Delta P = 2\gamma \cos \theta / R^{(1)}$. The pore volume fraction that is occupied by oil is called the oil saturation S_o.

$$S_o \equiv \text{Oil Saturation} = \frac{\text{Volume of Oil}}{\text{Pore Volume}} = 1 - S_w. \qquad \text{II.10}$$

S_w is the water saturation. It is assumed that oil and water are the only two phases present. As the pressure is again increased by ΔP, more oil enters the pores through still smaller apertures $R^{(2)}$, that is, $2\Delta P = 2\gamma \cos \theta / R^{(2)}$. By repeating this process through n increments ΔP, waiting for equilibrium to be attained, and then determining the volume of oil that has entered after each increase, a relation between oil pressure, $n\Delta P$ or equivalently $R^{(n)}$, and saturation S_o is obtained. This curve of oil pressure versus oil saturation is called the *drainage* curve.

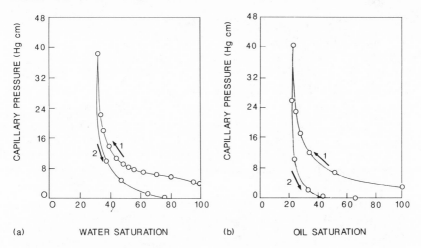

(a)   WATER SATURATION   (b)   OIL SATURATION

Fig. II.11. (*a*) Curves of drainage (1) and imbibition (2) for a sandstone. Water is the wetting fluid. (*b*) Curves of drainage (1) and imbibition (2) for a sandstone. Oil is the wetting fluid.

If this process is now reversed and the oil pressure progressively decreased, water (wetting phase) will flow back into the rock. A curve relating oil pressure versus oil saturation (*imbibition* curve) will be obtained which is different from the previously obtained drainage curve. A typical set of drainage-imbibition curves for consolidated sandstones under water-wet (*a*) and oil-wet (*b*) conditions are shown in figure II.11. Curve 1 in both cases represents the drainage curve of the wetting fluid. In both cases a pressure of approximately 4–5 cm of Hg was required for the nonwetting fluid to penetrate the largest pore throats and begin the displacement process. Because the pore size distribution is reasonably uniform, very small increases in pressure are sufficient to saturate the rocks to ~ 40% and ~ 30% wetting fluid saturations. At that point substantial increases in pressure resulted in only small further increases in saturation. The irreducible saturations at termination of drainage were ~ 37% and ~ 22%, respectively. This remaining saturation is presumably due to the bypassing of trapped water (fig. II.10) or the existence of microporous fraction within the pore structure that would require substantially larger pressures for oil penetration. There is a strong qualitative similarity between the set of two curves, one water-wet the other oil-wet, with the role of water and oil being interchanged. In both cases of imbibition (curve 2), when the capillary pressure was reduced from its maximum value to zero, there still remained a nonzero saturation of the nonwetting phase. This remaining volume fraction is called the "residual" saturation.

Capillary Pressure Curve and Pore Geometry

Determining the capillary pressure curve by mercury injection (mercury porosimetry) has become a standard method of characterizing the pore structure of rocks. The procedure consists of removing all pore water from the rock and then immersing it in a mercury bath at atmospheric pressure. Because the mercury does not imbibe spontaneously (the wetting fluid is air, contact angle $\theta = 140°$), the volume of mercury displaced equals the total rock volume. As the pressure is slowly increased, a small volume of mercury enters the rock. By continuously raising the pressure, the cumulative pore space occupied by mercury as a function of mercury pressure can be determined. Because the capillary pressure P_c is related to the smallest pore aperture R through which the mercury has to penetrate, $P_c = 2\gamma \cos \theta / R$, the capillary radius R can be used as the independent variable instead of the pressure P_c. An example of a mercury injection test by Pittman (1984) on the Nugget Sandstone is shown in figure II.12. The Nugget sandstone is a major reservoir in the overthrust region of Utah and Wyoming and is a rock composed of predominantly intergranular porosity. The photomicrograph clearly illustrates the intergranular nature of the porosity, but at this scale, does not readily show the cementing clay mineral illite. Note that the entry point for mercury is approximately $R = 50\mu$m. Between 50 and 10μm, the distribution of pores is uniform and represents 60% of the total porosity. At the limits of this test ($P = 122$ bars; $R = 0.06\mu$m) 94% of the pore space is saturated by mercury. The second example (fig. II.13) is the Baker Dolomite. Limestone was the original rock which was dolomitized leading to the development of pores among the dolomite rhombohedra. Mercury injection shows an entry radius of about 11μm, with uniform pore apertures between 11 and 4 μm accounting for approximately 50% of the pore volume. A hump, which is indicative of a bimodal pore size distribution, is also clearly seen. It represents micro-pores having radii in the range of 0.25 to 1μm and is clearly seen in the scanning electron micrographs of the rock (fig. II.13b). This sample has 20% porosity of which 12.5% is macroporosity (intergranular) and 7.5% microporosity (intragranular). The microporosity was formed within the original grains during dolomite formation.

Mercury porosimetry data can be used to quantify the structural parameters of a porous media in a simple fashion. Pore structure models can be subdivided into two broad categories. In the first category the pore structure is modeled as an array of parallel capillaries of different radii. Though topologically unrealistic, it is a very simple model to visualize and it captures many features of pore space structure. In the second category the porous medium is viewed as a network

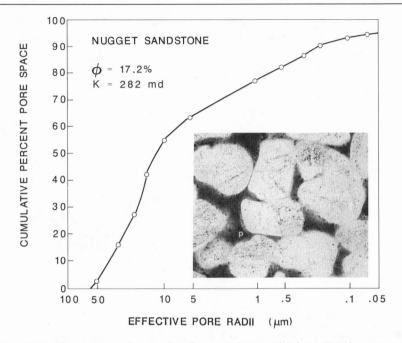

Fig. II.12. Mercury porosimetry data for a sandstone with intergranular porosity. The curve indicates a uniform distribution of pores with large radii. (After Pittman, E. D., 1984, The Pore Geometries of Reservoir Rocks, in *Physics and Chemistry of Porous Media* [Schlumberger-Doll Research, 1983, AIP Conference Proceedings, 107, Johnson, D. L. and P. N. Sen eds., N.Y.].)

of connected capillaries arising from the packing (random or ordered) of grains. The fundamental difference between the two models is that the network models have a multiply connected pore space. This leads to hysteresis in the drainage and imbibition curves as well as trapping of one fluid by the other.

From the mercury porosimetry curves, one can define a "pore volume" distribution function, more precisely, the pore volume fraction dV_p/V_p, which has entry radii in the range R to $R + dR$ for the nonwetting fluid.

$$dS = \frac{dV_p}{V_p} = -F(R)\,dR.$$ II.11

dS is the change in the nonwetting fluid saturation (mercury). Because capillary pressure is inversely proportional to the capillary radius, $dP/P = -dR/R$, we find:

$$dS = F(R)R\frac{dP}{P} = \frac{F(R)R^2}{2\gamma \cos \theta}\,dP.$$

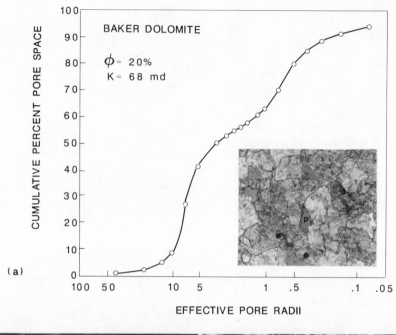

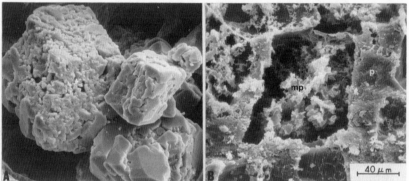

Fig. II.13. (*a*) Mercury injection curve indicating a bimodal pore size distribution (dolomite). (*b*) Micrographs taken with a scanning electron microscope: *A* —Dolomite rhombohedra with micropores; *B*—Pore cast showing large intercrystalline pores (*p*) and micropores (*mp*) within dolomite rhombohedron.

The distribution function $F(R)$ is obtained directly from the measured curves $S(R)$ or $S(P_c)$:

$$F(R) = -\frac{dS}{dR} = \frac{dS}{dP}\frac{2\gamma \cos \theta}{R^2}. \qquad \text{II.12}$$

$F(R)\,dR$ represents the fraction of the pore volume that is accessible to the nonwetting fluid through entry radii in the range R to $R + dR$. This result does not assume any specific geometric model for the pore space. As an illustration, we shall apply II.12 to the simple parallel capillary model discussed previously. Assume that the pore space can be described by an array of parallel capillaries of uniform cross-sections, of equal length L, but different radii R. Let $\psi(R)\,dR$ be the fraction of capillaries with radius between R and $R + dR$. At a pressure P_c, all capillaries of radii greater than $R = 2\gamma \cos \theta / P_c$ will be filled by the nonwetting fluid, while those of smaller radii will be filled with the wetting phase. The wetting phase saturation S_w can then be expressed as

$$S_w = \frac{\displaystyle\int_0^R \pi R^2 \psi(R)\,dR}{\displaystyle\int_0^\infty \pi R^2 \psi(R)\,dR} = \frac{\displaystyle\int_0^R R^2 \psi(R)\,dR}{\langle R^2 \rangle}$$

where $\langle R^2 \rangle$ is the second moment of the distribution $\psi(R)$. The examples illustrated by figures II.12 and II.13 give the mercury saturation $S_m = 1 - S_w$. Thus the experimental curves are directly related to the distribution function $\psi(R)$, as shown in the last equation. Differentiating this equation, we obtain

$$\psi(R) = \frac{\langle R^2 \rangle}{R^2} \frac{dS_w}{dR}.$$

In this way the distribution function $\psi(R)$ can be obtained from the experimentally determined curves $S_w(P)$ or $S_w(R)$. The unknown second moment $\langle R^2 \rangle$ is determined through the normalization condition on $\psi(R)$:

$$\int_0^R \psi(R)\,dR = 1.$$

A different interpretation of the mercury porosimetry curves has been presented by De Gennes and Guyon (1978). They draw an analogy between mercury injection and percolation (cf. chap. III). When mercury is injected at low pressure at a particular point, it forms a "multiply connected cluster" of finite size. Beyond a critical threshold pressure P_c, which corresponds to a critical pore throat radius R_c, the mercury forms a connected cluster which spans the rock. De Gennes and Guyon suggested that the porosimetry curve $S(R)$ in the vicinity of P_c is similar to the universal curve of percolation. That is, the saturation of mercury follows the law

$$S \sim (P - P_c)^\beta \qquad\qquad \text{II.13}$$

where β is a universal critical exponent. The saturation S is the analogue of the fraction of sites belonging to the infinite cluster at percolation (cf. chap. III, eq. III.10).

SURFACE ADSORPTION

Whenever there exists a concentration of solute atoms near a solid surface, there will be some adsorption. A simple physical picture of surface adsorption is that of a monolayer: a single layer of solute molecules that is adsorbed on the surface, all other molecules being considered in solution. This idealized layer interacts with the solid surface through molecular forces (Van der Waals forces) as in the case of physical adsorption, or through covalent bonds as in the case of chemical adsorption. Binding energies for these two cases are very different: the heat of physical adsorption is of the order of 4 kcal/mole, while the energy of chemical adsorption is on the order of 40 kcals/mole. By interacting with the molecules of a solid surface, the adsorbed monolayer lowers the surface energy of the solid-fluid interface. Because the adsorbed layer lowers the surface energy without changing the surface area, adsorption of molecules continues until the molecular chemical potential μ_s in the adsorbed layer is equal to the chemical potential μ_v in the bulk solution.

Variation of the Surface Tension: Gibbs' Equation

According to classical thermodynamics, the change in the internal energy and entropy of a system composed of N molecules adsorbed on a surface of area A is given by

$$dU = T\,dS + \mu\,dN + \gamma\,dA.$$

Here T is the temperature, μ is the chemical potential, and γ is the surface energy. The internal energy $U(S, N, A)$ is a linear homogeneous function of the extensive thermodynamic variables S, N, A:

$$U = TS + \mu N + \gamma A.$$

At constant temperature and pressure, an infinitesimal change in the internal energy can be expressed as

$$dU = T\,dS + \mu\,dN + N\,d\mu + \gamma\,dA + A\,d\gamma.$$

Comparing the two expressions for dU, we find:

$$A\,d\gamma = -N\,d\mu \; (P \text{ and } T \text{ constant}).$$

If we define $\Gamma = N/A$ as the surface concentration, we then arrive at Gibbs' equation which expresses the change in surface energy as the chemical potential changes:

$$d\gamma = -\Gamma d\mu.$$ II.14

Let us examine the simple example of a sandstone in contact with humid air. Considering the water in air as an ideal gas, the chemical potential of the water can be written as

$$\mu = \mu_o + RT \ln\left(\frac{P}{P_o}\right)$$

where R is the ideal gas constant, P the partial pressure of water, P_o the vapor pressure at saturation, and T the absolute temperature. The decrease $\Delta\gamma$ in surface energy between the two states, "sandstone in vacuum" and "sandstone in presence of saturated air," is

$$\Delta\gamma = -RT\int_0^{P_o} \Gamma \frac{dP}{P}.$$

Numerically one obtains $\Delta\gamma \approx 300$ mJ/m^2 (Murphy et al., 1984). The adsorption of several monolayers of water has important consequences on the acoustic properties of rocks. For example, the shear wave velocity of a Coconino sandstone decreases from 2.6 km/sec to 2.3 km/sec as P increases from 0 to $0.8P_o$ (Clark et al., 1984). This result can be interpreted by noting that the elastic modulus M varies with the surface energy γ as $M \sim \gamma^{1/3}$, and that γ varies with (P/P_o) as shown in the last equation. The relation between M and γ arises due to the fact that the force of interaction F between grains is directly proportional to γ, and elastic contact theory implies that $M \sim F^{1/3}$.

Charged Double Layer

Consider the example of a quartz crystal, in the presence of water or acid, which has just been fractured. These "fresh" surfaces created by a fracture present highly unstable structures. The broken Si-O covalent bonds are electrically charged, dangling bonds. Very low water-vapor pressures are sufficient to hydroxylate the bonds and diminish the surface energy. Surface hydroxyls are dominated by the groups Si-OH (silanol group) (fig. II.14a, b). Let SOH$_{(surf)}$ denote a surface hydroxyl site. The adsorbed water vapor lowers the surface energy, as described by Gibbs' adsorption equation (fig. II.14c). Assume that the surface hydroxyls and adsorbed water are in contact with the liquid water. Chemical equilibrium for this process is described by:

$$H^+ + SOH = SHO_2^+ \qquad K_1 = \frac{[SOH_2^+]}{[H^+][SOH]}$$

$$SOH = SO^- + H^+ \qquad K_2 = \frac{[H^+][SO^-]}{[SOH]}.$$

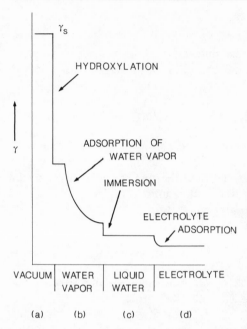

Fig. II.14. Evolution of a "freshly" fractured quartz surface. *A*—fresh surface; *B*—hydroxylated surface; *C*—adsorbed surface; *D*—wetted surface. (After Parks, G. A. 1984, Surface and interfacial free energies of quartz, *J. Geophys. Res.* 89:3997–4008.)

This balance results in an electrical surface charge σ_o:

$$\sigma_o = (\Gamma_{SOH_2^+} - \Gamma_{SO^-})e$$

where Γ represents the surface concentration of the ionic species and e the electron charge. One defines the point of zero charge by

$$[SOH_2^+] = [SO^-]$$

with

$$\frac{K_2}{K_1} = [H^+]^2 \frac{[SO^-]}{[SOH_2^+]}.$$

At this point the pH is: $pH = -\dfrac{1}{2}\ln\left(\dfrac{K_2}{K_1}\right) \cong 3$ (Parks, 1984). When the pH > 3 the surface charge σ_o is negative. On the other hand, when pH < 3, σ_o is positive.

When $\sigma_o \neq 0$, the surface electrical charge is compensated by a redistribution of ions in solution to establish electro-neutrality. There exists a diffuse layer of charge which can be thought of as a superficial layer of thickness h and equivalent surface charge density σ_d, such that $\sigma_o + \sigma_d = 0$ (fig. II.15). This is the electrical "double-layer" model of Gouy-Chapman. Take for example, quartz in contact with saline water with a pH greater than 3:

$$\Gamma_{H^+} + \Gamma_{Na^+} = \Gamma_{OH^-} + \Gamma_{Cl^-} \qquad \text{with } \Gamma_{OH} \gg \Gamma_{H^+} \text{ and } \Gamma_{Na^+} \gg \Gamma_{Cl^-}.$$

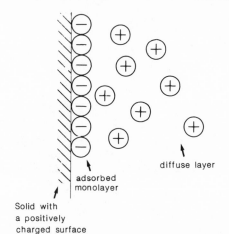

diffuse layer

adsorbed
monolayer

Solid with
a positively
charged surface

Fig. II.15. Model of the electrical
double layer. Case of a positive
surface charge.

The variation in surface energy is given by $d\gamma = -\sum_{i=1}^{4} \Gamma_i \, d\mu_i$ where the
index i corresponds to the four species H^+, Na^+, OH^-, and Cl^-. If the
pH is constant, μ_{H^+} and μ_{OH^-} are constant. If further Γ_{Cl^-} is negligible, we find

$$\Delta\gamma \cong -2.3RT\Gamma_{Na^+}\Delta \ln[Na^+] \approx 20 \text{ mJ m}^{-2}.$$

The observed variation in energy is small.

SURFACE ROUGHNESS

Just as the parameter ϕ does not give a complete description of all
aspects of porosity, the parameter S (specific surface) does not give a
complete description of the internal pore surface. We have seen that to
a first approximation, S is inversely proportional to the grain diameters.
In reality grains are often not spherical, their surfaces not smooth, and
they possess a complex structure of their own. A description of this
structure is very important for a number of properties: mechanical,
electrical, and dielectric in particular.

Models of Asperities

Figure II.16 shows schematically the topography of an internal surface.
Pore surface topography is in general quite irregular. It can be characterized through a distribution of asperity heights, with the distribution
of asperity maxima being of particular importance for mechanical
properties. When two rough surfaces approach each other (as is the
case when a confining pressure is applied), the asperities come into

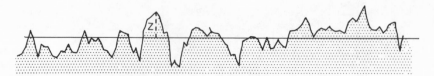

Fig. II.16. Distribution of asperities on a pore surface.

contact only at discrete points along the surface. This is in marked contrast to a continuous contact between smooth surfaces. Pore compressibility (closure) to a large extent is controlled by the asperity distribution on grain surfaces and their interaction at grain contacts.

Notion of Fractal Dimension

The distribution of asperities on a pore surface is such that in reality this "surface" is not actually a surface, but an object in two dimensions. The concept of a fractal allows a convenient method for describing these types of "surfaces" or "curves."

The notion of fractal dimension can be introduced in the following simple fashion. Suppose a reference volume V_o is composed of N smaller elementary volumes r^D, that is, $N = V_o/r^D$. The smaller volume r^D is the reference volume used for measurement. For example, with $D = 1$, a segment of unit length can be decomposed into N smaller segments of length r, $N = 1/r$. In two dimensions ($D = 2$), a surface of unit area can be decomposed into N smaller areas $1/r^2$. Generalizing this procedure, a dimension D can be defined through the relation (Mandelbrot, 1983)

$$D = \frac{\log N}{\log(1/r)}.$$ II.15

This definition allows the notion of a noninteger dimension, or fractional dimension. N is the number of elementary elements necessary to cover the unit surface, curve, or volume. An equivalent formulation of this result is that $r = N^{-(1/D)}$ is the similarity variable between the elementary element and the whole. Thus, in the case where the fractal curve has length L when measured with a ruler of length ε, $D = \frac{\log(L/\varepsilon)}{\log(1/\varepsilon)}$. Solving for L

$$L(\varepsilon) = \varepsilon^{1-D}.$$ II.16

If $D = 1$ (Euclidian dimension), L is a constant independent of ε. When $D \neq 1$, the length of the curve depends on the choice of ε. An example of this type of a curve is Von Koch's curve (fig. II.17) which is

Fig. II.17. Von Koch's curve. The segment *AB* is replaced by 4 segments each of length *AB*/3. This operation is repeated *n* fold.

obtained by a process of repeated dissection. A segment *AB* is dissected into four new segments, each being one-third the original length. This operation is repeated at the next stage: each of the four segments obtained on the previous step is dissected into four new segments each being one-third the length at the preceding step, and so forth. If ε is the length of the elementary segment at stage (*n*), then at stage (*n* + 1), the elementary segment will be of length $\varepsilon/3$. Let $L(\varepsilon)$ represent the total length at stage (*n*) and $L(\varepsilon/3)$ the total length at stage (*n* + 1). This implies:

$$L(\varepsilon/3) = \frac{4}{3}L(\varepsilon).$$

This result is a particular example of the general relation II.16. If $L(\varepsilon) = \varepsilon^{1-D}$, then one obtains

$$(\varepsilon/3)^{1-D} = \frac{4}{3}(\varepsilon)^{1-D}$$

with

$$D = \frac{\log 4}{\log 3} = 1.26.$$

Von Koch's "curve" is a fractal of dimension 1.26. It is a self-similar curve, that is, a curve invariant on a change of scale. The ratio of self-similarity is 4/3. The curve is continuous but not differentiable, and its length is infinite in the limit $\varepsilon \to 0$.

Natural Fractal Pore Surfaces

Katz and Thompson have shown (fig. II.18) that the pore space of certain sandstones is a fractal quantity, being self-similar over three to

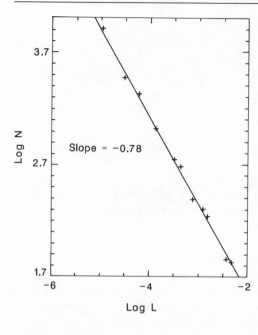

Fig. II.18. Log-Log plot of the number of microstructural elements (per unit length) of size L, as a function of the size L (in cm) for Coconino sandstone. The slope of the curve is equal to 2-D. (After Katz, A. J., and A. H. Thompson, 1985, Fractal Sandstone pores: Implication for conductivity and pore formation, *Phys. Rev. Letters* 54:1325–28.)

four orders of magnitude. Their measurements, with the aid of optically and electronically obtained micrographs, covered the range 100 Å–100 μm. The values of D obtained by Katz and Thompson ranged from 2.57 to 2.87.

How does one explain the fractal nature of pores as indicated by this evidence? The explanation must lie in the rock history and, very probably, in the processes acting during diagenesis (crystallization-dissolution). The observed results indicate that in most cases D is variable, and the mechanisms responsible for the fractal nature do not give a unique value of D, but rather a range of values.

PROBLEMS

II.a. Minimum Porosity of an Assembly of Identical Spheres

A hexagonal close packing is formed by a sequence $ABAB\ldots$ of compact layers. The elementary cell is defined by three vectors $\vec{a}, \vec{b}, \vec{c}$. $\vec{a}$ and $\vec{b}$ are in the basal plane, are of equal length, and form an angle of 120°. $\vec{c}$ is perpendicular to the basal plane.

(1) Show that the ratio c/a is 1.630.
(2) Calculate the porosity of this hexagonal packing. Compare this with the porosity of a random packing of spheres.

II.b. Surface Energy and Young's Modulus

Orowan's model expresses the elastic force σ (per unit surface area) exerted by an interatomic bond in terms of the atomic displacement x (as measured from equilibrium position), $\sigma = \sigma_o \sin\left(2\pi\dfrac{x}{b}\right)$. Here b is the interatomic distance and σ_o the maximum force the bond can support.

(1) Young's modulus E (cf. chap. V) is defined by the relation $\sigma = E\varepsilon$, where $\varepsilon = \dfrac{x}{b}$, valid in the range $x \ll b$. Calculate E as a function of σ_o.

(2) The surface energy γ is the energy required to form a surface of unit area. Show that $\gamma \cong \dfrac{Eb}{4\pi^2}$.

II.c. Gouy-Chapman Model of the Electrical Double Layer

When a solution containing cations and anions of equal and opposite charges $\pm ze$ is in contact with a positively charged solid surface, it develops an electrical "double layer." Assume that the surface is planar and that the electrical potential V only depends on the coordinate x normal to the surface.

(1) Assuming that the concentration of ions in solution is given by Boltzmann's law, $n^- = n_o\exp\left(\dfrac{zeV}{kT}\right)$ and $n^+ = n_o\exp\left(-\dfrac{zeV}{kT}\right)$, show that the potential V is a solution of the equation

$$\frac{d^2V}{dx^2} = \frac{2\,ze\,n_o}{\varepsilon\,\varepsilon_o}\sinh\left(\frac{zeV}{kT}\right).$$

Utilize the relations of electrostatics $\vec{\nabla}\cdot\vec{E} = \dfrac{\rho}{\varepsilon\,\varepsilon_o}$ and $\vec{E} = -\vec{\nabla}V$, where ρ is the charge density.

(2) If the thermal energy is much larger than the electrostatic, $zeV \ll kT$, show that $V = V_o\exp(-x/\lambda)$ and calculate the critical distance λ for a solution of 10^{-1} moles/liter and 10^{-5} moles/liter. Assume $z = 1$ and $\varepsilon = 80$.

$$k = 1.38 \times 10^{-23}\text{J}/°\text{K} \quad \text{(Boltzmann's constant)}$$

$$e = 1.60 \times 10^{-19}\text{C} \quad \text{(electron charge)}$$

$$\varepsilon_o = 8.9 \times 10^{-12} \quad \text{(permittivity of a vacuum)}$$

II.d. Adsorption of a Monolayer (Langmuir Isotherm)

In certain cases the specific surface of a porous medium can be measured by the volume of gas that is adsorbed on the pore surfaces. The curves of adsorbed volume as a function of pressure P, $V_{ads} = f(P)$, are called the adsorption isotherms. Let S denote the total available pore surface, S' the part adsorbed, and S'' the free surface; $S = S' + S''$.

(1) Assuming that the rate of evaporation of an adsorbed species is proportional to S' and the rate of condensation to P and S'', show that $\dfrac{S'}{S} = \dfrac{bP}{1 + bP}$.

(2) Assuming that the adsorption is restricted to a monolayer, show that the adsorbed volume V follows the law: $\dfrac{P}{V} = \dfrac{1}{bV_m} + \dfrac{P}{V_m}$. ($V_m$ is the adsorbed volume corresponding to complete coverage by a monolayer).

REFERENCES

Adamson, A. W. 1982. *Physical Chemistry of Surfaces*. New York: Wiley.

ASTM. 1964. *Procedures for Testing Soils*, 4th edition. Philadelphia: American Society for Testing and Materials.

Beard, D. C., and P. K. Weyl. 1973. Influence of Texture on Porosity and Permeability of Unconsolidated Sand. *AAPG Bulletin* 57:349–369.

Clark, V. A., B. R. Tittman, and T. W. Spencer. 1980. Effects of Volatiles on Attenuation and Velocity in Sedimentary Rocks. *J. Geophys. Res.* 35:5190–98.

Clarke, R. H. 1979. Reservoir Properties of Conglomerates and Conglomeratic Sandstones. *AAPG Bulletin* 63:799–809.

Craig, F. F., Jr. 1971. *The Reservoir Engineering Aspects of Water-Flooding*. Society of Petroleum Engineers of AIME, Monograph 3. Dallas, Texas.

Defay, R., and I. Prigogine (with collaboration of A. Bellemans). 1966. *Surface Tension and Adsorption* (translated by D. H. Everett). London: Longmans Green.

De Gennes, P. G., and E. Guyon. 1978. Lois generales pour l'injection d'un fluide dans un milieu poreux aleatoire. *Journal de Mecanique* 17:403–32.

Mandelbrot, B. B. 1983. *The Fractal Geometry of Nature*. New York: Freeman.

Murphy, W. F., K. W. Winkler, and R. L. Kleinberg. 1984. Frame Modulus Reduction in Sedimentary Rocks: the Effect of Adsorption on Grain Contacts. *Geophys. Research Letters* 11:805–8.

Palciauskas, V. V., and P. A. Domenico. 1989. Fluid Pressures in Deforming Porous Rocks. *Water Resources Research* 25:203–13.

Scheidegger, A. E. 1974. *The Physics of Flow Through Porous Media*. 3d ed. Toronto: Univ. of Toronto Press.

Terzagi, K. 1936. The Shearing Resistance of Saturated Soils. *Proc. 1st Int. Conf. Soil Mech., Cambridge, Mass.* 1:54–56.

III. Heterogeneous Media

ROCKS ARE inherently heterogeneous materials: in some cases only slightly heterogeneous, while in others, extremely so. A single rock can be considered both slightly or highly heterogeneous, depending on what scale is being considered and which property is being studied. The notion of scale is an important concept that appears frequently in the earth sciences. When the heterogeneity is not too strong and results primarily from disorder at a small (micro) scale, it is useful to define several other (larger) scales: mini and macro (completing the "3 Ms"). When the heterogeneity is very strong this approach is not sufficient, because in a certain sense all scales are similar. In these situations the concept of percolation, which embodies the principal of scale invariance, becomes the central concept. Percolation theory, which has seen remarkable developments in recent years, is not a commonly used tool in the Earth Sciences, yet it is very appropriate for the analysis of complex and sometimes novel properties of strongly heterogeneous systems.

MICRO, MINI, MACRO SCALES

Micro- and Minivariables

Introducing the three scales, micro, mini, and macro—the 3 Ms—allows the real medium to be described in the following manner. At the micro-scale, one views the elementary rock constituents, grains and pores. At the mini-scale, one observes a volume which contains a large number of grains and pores. The macro-scale is defined by the scale of the rock sample being studied (fig. III.1). By definition

$$\text{Micro} \ll \text{Mini} \ll \text{Macro}.$$

Microvariables are the variables at the micro or grain scale:

(*a*) microstructural parameters such as grain and pore diameters, fractures, etc.

(*b*) physical parameters such as stresses, displacements, etc. at the micro-scale.

Minivariables are averages of the grain scale physical variables. The effective properties of the medium are defined and calculated from the minivariables. Consider for example the vector displacement of a point

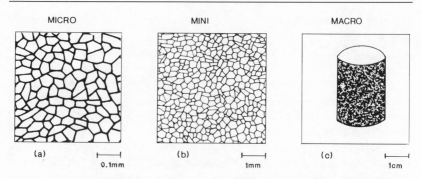

MICRO   MINI   MACRO

(a)   (b)   (c)
├────┤   ├────┤   ├────┤
0.1mm   1mm   1cm

Fig. III.1. The 3 scales: Micro, Mini, and Macro.

within a medium. One defines $\bar{u}_i(\vec{x})$, the average displacement in the effective medium, from the displacement u_i averaged over a representative volume element V_R:

$$\text{Micro} \ll V_R \ll \text{Macro} \quad \bar{u}_i = \frac{1}{V_R} \iiint_{V_R} u_i(x, y, z)\, dx\, dy\, dz.$$

By replacing the real discontinuous medium by an effective continuous one, it is possible to describe the variation of $\bar{u}_i$ on a larger (macro) scale (fig. III.2). Similarly, one can define in V_R an average temperature $\bar{T}$, an average electrical potential $\bar{V}$, an average concentration $\bar{c}$, and so on. The physical response of the medium is described in terms of *effective properties*: these are the average properties of a representative volume element V_R. This approach is applicable to cases where the medium is statistically homogeneous, that is, it is heterogeneous on a small scale but appears homogeneous on a large scale. Statistical homogeneity is the basic condition for the application of the effective medium approach.

Effective properties of a medium can also be defined utilizing an appropriate expression for the energy (elastic energy, thermal energy,

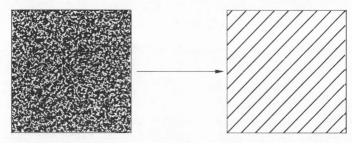

Fig. III.2. Real heterogeneous medium and effective homogeneous medium.

etc.). The two approaches have been shown to be equivalent (Hashin, 1983).

The Classical Approximation

Consider the elastic properties of an ideal rock, which is assumed to be composed of a random packing of identical grains. Such a rock is statistically homogeneous. Let σ be the applied confining stress and ε the deformation of the grain in response to σ. Anticipating chapter IV, we know that ε is a linear function of σ, $\sigma = M\varepsilon$, where M is an elastic modulus of the grain material. The modulus M could be the incompressibility K or the shear modulus μ (chap. IV). Suppose that a stress field is applied to this rock and that the stress field is statistically homogeneous on the macro scale. In this case the determination of the effective elastic moduli is simple:

$$M^* = \frac{\overline{\sigma}}{\overline{\varepsilon}}$$

where $\overline{\sigma}$ and $\overline{\varepsilon}$ are averages over V_R, the representative volume element.

However, it frequently occurs that the stresses, deformations, and other physical variables are not statistically homogeneous. The medium, from the point of view of composition, structure, etc., is statistically homogeneous, but the constraining field is not (it varies from point to point on the macro scale). What is the effective elastic modulus for this case? For this situation there is no longer a local linear relationship between $\overline{\sigma}(\vec{x})$ and $\overline{\varepsilon}(\vec{x})$ because the stress at $\vec{x}$ depends on deformations at all points $\vec{x}'$ of the medium and vice versa. The theory is non-local. By approximation and development in series, one obtains for a one-dimensional model:

$$\overline{\sigma}(x) = M_o^* \varepsilon(x) + M_1^* \frac{\partial \overline{\varepsilon}}{\partial x}(x) + M_2^* \frac{\partial^2 \overline{\varepsilon}}{\partial x^2}(x) + \cdots$$

If the stress and strain gradients are small, the derivatives of $\overline{\varepsilon}$ can be neglected and one recovers the previous result.

$$M^* = M_o^* = \frac{\overline{\sigma}}{\overline{\varepsilon}}.$$

This is the classical approximation (Beran and McCoy, 1970). The classical approximation is adequate for the majority of statistical problems. It is not true for dynamic problems, especially at high frequencies, when the deformation gradients are large and wavelengths are very small. In what follows we limit ourselves to the classical approximation.

Table III.1. ANALOGIES BETWEEN ELASTICITY, THERMAL CONDUCTION, AND ELECTRICAL CONDUCTION

Elasticity	Thermal Conduction	Electrical Conduction
displacement u_i	temperature T	potential V
deformation	gradient $\dfrac{\partial T}{\partial x_i}$	electric field $E_i = -\dfrac{\partial V}{\partial x_i}$
$\varepsilon_{ij} = \dfrac{1}{2}\left(\dfrac{\partial u_i}{\partial x_j} + \dfrac{\partial u_j}{\partial x_i}\right)$		
stress σ_{ij}	heat flux J_i	flux of current j_i
elastic moduli C_{ijkl}	thermal conductivity λ_{ij}	electrical conductivity C_{ij}

CALCULATION OF EFFECTIVE PROPERTIES

Analogies between Different Physical Properties

We are interested in the mechanical, electrical, dielectric, magnetic, and thermal properties of rocks, each of which will be the subject of a subsequent chapter. There exists a formal mathematical relationship between the physical properties in these different domains. This analogy is quite useful because a calculation of an effective property in one domain can be directly carried over to another. Table III.1 presents the relationships between variables in the domains of elasticity, thermal conductivity, and electrical conduction. This analogy between properties is almost exact because for each of the three properties the constitutive relation is described by a linear relationship. Avoiding tensorial notation, we list these constitutive relationships in one-dimension:

$\sigma = M\varepsilon$:   Stress is proportional to the displacement (Hooke's Law).

$J_x = -\lambda\dfrac{\partial T}{\partial x}$:   Heat flux is proportional to the temperature gradient (Fourier's Law).

$j_x = CE_x$:   The electric current is proportional to the electric field (Ohm's Law).

In all three cases, the response is proportional to the applied force, and the medium properties are characterized through the coefficients of proportionality:

$$M, \lambda, C.$$

Below we shall discuss the example of elasticity in more detail, but the results obtained here can be carried over to the computation of effective electrical and thermal properties. Elasticity is somewhat more complex mathematically because the variables σ_{ij} and ε_{ij} are always second-rank tensor properties. This implies that the moduli M are tensors of rank 4 (chap. IV). On the other hand, temperature gradients, electric fields, heat fluxes, and electric currents are all vector quantities, implying that thermal and electric conductivities are second-rank tensors (chaps. VIII and X). For each of these three cases, at equilibrium or at steady state, the stress σ_{ij} or fluxes J_i, j_i obey the same conservative equation:

$$\frac{\partial J_i}{\partial x_i} = 0.$$

This equation expresses mathematically the conservation of energy, electric charge, or stress equilibrium when external sources are absent.

The analogies discussed above can be extended to other properties that will be examined later in this book: permeability (chap. V), dielectric permittivity (chap. IX), and magnetic permeability (chap. XI). Linear laws corresponding to these properties are as follows:

$q_x = -\dfrac{k}{\eta}\dfrac{dp}{dx}$: The volume flux of fluid q_x is proportional to the pressure gradient dp/dx: k is the permeability and η is the fluid viscosity (Darcy's law).

$D_x = \varepsilon E_x$: Electrical induction D_x is proportional to the electric field E_x: ε is the dielectric permittivity.

$B_x = \mu H_x$: Magnetic induction B_x is proportional to the magnetic field H_x: μ is the magnetic permeability.

Direct Calculation

Conceptually the simplest method of calculating effective properties is direct calculation. We can illustrate this method by computing the elastic properties of a medium which is composed of two isotropic phases, each of which is elastic. This could be a model for a monomineralic rock (first phase), which contains a certain amount of pore space saturated with fluid (second phase). Let v_1 and v_2 represent the volume fractions of the two phases and M_1 and M_2 be their elastic moduli. The average stress and strain are then

$$\bar{\sigma} = \bar{\sigma}^{(1)} v_1 + \bar{\sigma}^{(2)} v_2 \quad \text{and} \quad \bar{\varepsilon} = \bar{\varepsilon}^{(1)} v_1 + \bar{\varepsilon}^{(2)} v_2.$$

Here the indices (1) and (2) represent the phase. The definition of an effective modulus M^* implies the relations:

$$M^* \equiv \frac{\bar{\sigma}}{\bar{\varepsilon}} = \frac{\bar{\sigma}^{(1)}}{\bar{\varepsilon}} v_1 + \frac{\bar{\sigma}^{(2)}}{\bar{\varepsilon}} v_2.$$

On introducing the moduli M_1 and M_2, we find:

$$M^* = M_1 \frac{\overline{\varepsilon}^{(1)}}{\overline{\varepsilon}} v_1 + M_2 \frac{\overline{\varepsilon}^{(2)}}{\overline{\varepsilon}} v_2.$$

Because $\overline{\varepsilon}^{(1)} v_1 = \overline{\varepsilon} - \overline{\varepsilon}^{(2)} v_2$ we find

$$M^* = M_1 + (M_2 - M_1) \frac{\overline{\varepsilon}^{(2)}}{\overline{\varepsilon}} v_2. \qquad \text{III.1}$$

This relation can be used for both the modulus of incompressibility K and shear modulus μ. To complete the calculation of M^*, the term $\dfrac{\overline{\varepsilon}^{(2)}}{\overline{\varepsilon}}$ must be evaluated. This in turn requires that a precise geometric model for the distribution of phases be specified.

In the simple case of a small concentration of spherical pores ($v_2 \ll 1$) imbedded in phase 1, one can neglect the interactions between pores and treat them individually. In this case one can easily calculate the volume strain $\dfrac{\overline{\varepsilon}^{(2)}}{\overline{\varepsilon}}$ and obtain the incompressibility modulus K (see Landau and Lifshitz, 1967, chap. IV):

$$\frac{\overline{\varepsilon}^{(2)}}{\overline{\varepsilon}} = \frac{(3K_1 + 4\mu_1)}{(3K_2 + 4\mu_1)}. \qquad \text{III.2}$$

Combining equations (1) and (2) and assuming the pores are empty, $K_2 = 0$, $v_2 = \phi$, we find

$$K^* = K_1 \left[1 - \left(1 + \frac{3K_1}{4\mu_1} \right) \phi \right]. \qquad \text{III.3}$$

K^* is the "effective" bulk modulus of this porous medium. It depends on the two moduli, K_1 and μ_1, and on ϕ (fig. III.3). For more complex cases (more than two phases, large concentrations, nonspherical geometries, etc.) the direct method of calculation loses its simplicity and becomes impractical.

Many interesting microstructural models are based on an assemblage of composite spheres. These composite spheres of radius R_1 contain internal spheres of radius R_2 with material properties K_2 and μ_2, and an outer shell of material properties K_1 and μ_1. The ratio R_2/R_1 is related to the volume fraction of medium 2, $(R_2/R_1)^3 = v_2$. Each of these composite spheres has an effective modulus K^*, which is the modulus that an equivalent homogeneous sphere of radius R_1 would have. The effective modulus for this model is:

$$K^* = K_1 + (K_2 - K_1) \frac{(3K_1 + 4\mu_1)v_2}{(3K_2 + 4\mu_1) - 3(K_2 - K_1)v_2}. \qquad \text{III.4}$$

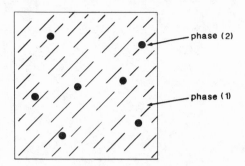

Fig. III.3. Heterogeneous medium of 2 components, with $v_2 \ll v_1$.

A medium which contains no void space can be formed with such composite spheres ($R_2/R_1 =$ constant). To accomplish this there must be a distribution in the composite sphere sizes which ranges to the infinitesimal, $R_1 \to 0$. On a scale much larger than the sphere radii, such a medium of composite spheres then appears homogeneous with effective modulus K^* given by III.4.

Equation III.4 will be used in chapters VIII and X for computing theoretical bounds on the electrical and thermal conductivity of an isotropic medium (the Hashin-Shtrikman bounds). This will be accomplished by replacing the incompressibility K by the conductivity λ or C and utilizing the formal analogies between properties discussed previously in this chapter.

Upper and Lower Bounds

A different approach for estimating effective properties like M^* consists of bounding M^* by an upper and lower value. As an example, consider a new medium composed of two isotropic phases. Suppose that the phases are distributed in parallel layers (beds), stacked alternately, with volume fractions v_1 and v_2, respectively (fig. III.4). For this medium there are two values which limit the effective elastic modulus: one corresponds to a deformation perpendicular to the beds (case a) and the other parallel to the beds (case b):

$$M_a = \left(\frac{\overline{\sigma}}{\overline{\varepsilon}}\right)_a \quad M_b = \left(\frac{\overline{\sigma}}{\overline{\varepsilon}}\right)_b.$$

Case (a) implies that the stress is uniform and equal in each bed:

$$\overline{\sigma} = \overline{\sigma}^{(1)} = \overline{\sigma}^{(2)}, \quad \overline{\varepsilon} = \overline{\varepsilon}^{(1)} v_1 + \overline{\varepsilon}^{(2)} v_2.$$

Case (b) implies that the deformation (strain) in each bed is the same:

$$\overline{\sigma} = \overline{\sigma}^{(1)} v_1 + \overline{\sigma}^{(2)} v_2, \quad \overline{\varepsilon} = \overline{\varepsilon}^{(1)} = \overline{\varepsilon}^{(2)}.$$

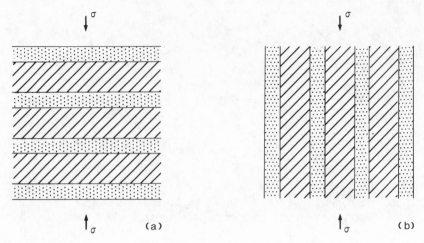

Fig. III.4. Medium composed of layers. a—stress applied perpendicular to layers; b—stress applied parallel to layers.

As a result $\dfrac{1}{M_a} = \dfrac{v_1}{M_1} + \dfrac{v_2}{M_2}$ and $M_b = v_1 M_1 + v_2 M_2$, where

$$M_a \leq M^* \leq M_b.$$

This result can be generalized to n isotropic phases

$$\frac{1}{M_a} = \sum_{i=1}^{n} \frac{v_i}{M_i} \quad \text{and} \quad \frac{1}{M_b} = \sum_{i=1}^{n} v_i M_i. \qquad \text{III.5}$$

The main interest of these bounds is their simplicity. But if the two phases M_1 and M_2 are significantly different, then M_a and M_b are also significantly different. In these cases more information is required so as to locate where M^* lies between the two bounds. Suppose for example $M_1 = 10^3 M_2$ and $v_1 = v_2 = 1/2$. Then $M_a \sim 2M_2$ and $M_b \sim 500M_2$! This example clearly shows the limitations of this method of bounds.

If in addition we know that a composite material is macroscopically isotropic, then improved bounds can be obtained by considering models of assemblages of composite spheres. By assuming that a sphere of material (1) contains a sphere of material (2), with $(R_1/R_2)^3 = v_2$, one obtains the first bound (eq. III.4). The other bound is obtained by interchanging the materials: sphere of (2) contains sphere of (1), with $(R_2/R_1)^3 = v_1$. The bounds are known as the Hashin-Shtrikman bounds (Hashin, 1983).

Self-Consistent Method

A third method for determining effective properties consists of introducing a self-consistency approximation into the direct calculation. We have seen that a direct calculation is only possible for simple cases; for example equation III.1 yields the result for M^*, if one can compute $\dfrac{\bar{\varepsilon}^{(2)}}{\bar{\varepsilon}}$. But this is precisely the difficult part of the computation, especially when the volume fractions of (2) are not small and the interactions between particles (2) are not negligible. Self-consistent methods attempt to take into account these complexities. The self-consistent method can be illustrated through the following example. Assume that a small spherical amount of phase (2) is imbedded into a medium which has the properties of the effective medium (moduli K^* and μ^*) rather than medium (1). In this case $\dfrac{\bar{\varepsilon}^{(2)}}{\bar{\varepsilon}}$ is given by III.2:

$$\frac{\bar{\varepsilon}^{(2)}}{\bar{\varepsilon}} = \frac{3K^* + 4\mu^*}{3K_2 + 4\mu^*}. \qquad \text{III.6}$$

By considering that phase (2) is imbedded into the effective medium instead of phase (1), one takes into account the interactions between phase (2) particles. Thus III.6 should be a better approximation than III.2 for large values of v_2. Equation III.1 now yields two equations, one for K^* and the other for μ^*, which can be solved self-consistently for K^* and μ^*.

This self-consistent approach incorporates the principle of the three scales, the "3Ms." We identify the micro- (K_1, μ_1) and minivariables (K^*, μ^*), and then view the medium on a larger scale where it appears that phase 2 is being imbedded into the effective medium rather than phase 1.

PERCOLATION

Effective medium description becomes inadequate when the medium considered is strongly heterogeneous. Near the critical concentration for a percolation threshold, there occur effects which cannot be anticipated through effective medium computations. The permeability, electrical conductivity, and elastic moduli approach zero near the percolation threshold.

Bond and Site Percolation on a Network

Consider a square network in two dimensions. Each of these small squares may be occupied randomly with probability p. We define p as

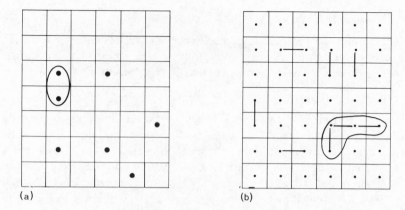

Fig. III.5. Site and bond percolation.

the probability that a site is occupied. Two squares are considered nearest neighbors if they have a common side. When two (or more) adjacent squares are occupied, these squares form a connected cluster (fig. III.5a).

Let us assume that all squares are occupied by spheres. One can connect two nearest neighbor spheres by a line, which we will call a bond. Let p be the probability that two spheres are connected by a bond. A cluster is formed when many nearest neighbor sites are connected by bonds (fig. III.5b). The existence of a bond can be interpreted, for example, as the ability of electric current or fluid to flow between the connected sites.

When p increases, the number of occupied sites or bonds increases and the clusters increase in size. At a certain critical value $p = p_c$, an "infinite" connected cluster is formed. That is, there exists a connected cluster which spans the network from one side to the other. This percolation threshold depends on the type of network (square, triangular), dimension, and type of percolation (site or bond). Values of the critical probability p_c for several networks are listed in table III.2 (Stauffer, 1985).

Table III.2. CRITICAL PROBABILITY FOR PERCOLATION

Network	Site	Bond
Square	0.59	0.50
Triangular	0.50	0.34
Cubic	0.31	0.24

It is useful to know certain quantities, such as the average size S of a cluster and the fraction of sites P that belong to the infinite cluster, because they can be related to physical properties. Below the threshold, there exist finite clusters of variable sizes, the average size being S. For example, in an insulating medium all conducting clusters would be isolated. S is approximately equal to the correlation length ξ, with ξ being a measure of the average distance between two sites that belong to a connected cluster. As $p \to p_c$, S and ξ approach infinity. This explains why an effective medium description becomes inadequate in this limit. As one gradually approaches p_c, the scale over which we must carry out an average (the representative volume element), becomes larger and larger: it is no longer possible to define a scale "mini" in the vicinity of p_c. All scales become equivalent. In the vicinity of the threshold, the cluster is a fractal, self-similar object (Mandelbrot, 1982), with fractal dimension of approximately 2.5 in three dimensions. Close to the percolation threshold, the correlation length ξ has the form

$$\xi \approx (p - p_c)^{-x}$$

where x is a critical exponent which is independent of the network geometry. In three dimensions $x \approx 0.88$.

The quantity P, fraction of sites in the infinite cluster (above the threshold), is also a physically relevant quantity. For example, the network conductivity is related to P.

Bethe Lattice

It is generally not easy to compute the previously discussed quantities S and P, and for most networks there do not exist exact solutions for these quantities. However, there is one network for which the problem allows an exact solution. This is the Bethe lattice (fig. III.6) where each site is connected to z neighbors. z is called the network coordination number.

To construct a Bethe network, pick an original site which will be denoted by O. This site is connected to four adjacent sites A (coordination $z = 4$). From every site A there are four bonds: one connected to the original site O and $(z - 1) = 3$ to new sites B. Each of the new sites B is connected to $(z - 1) = 3$ new sites C and so on. By continuously repeating this branching process, an infinite network is constructed. Note that a return to the original site O by a different path is not possible. The Bethe lattice has no closed paths.

Now assume that on this network a site or a bond is occupied with a probability p. What is the percolation threshold p_c for this network? The answer can be obtained by examining how an infinite path is

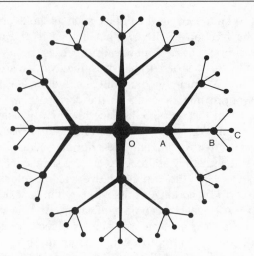

Fig. III.6. Bethe network with $z = 4$.

constructed. Starting from the original site O, there are z possible paths
to reach an A site. Each of the paths leaving A, and not returning to
O, reach a site B which is occupied with a probability p. To form an
infinite path, a return to any previous site is not allowed. In an infinite
Bethe lattice, at every step in the process there will be $(z - 1)$ possible
new sites, each being occupied with probability p: thus $[(z - 1)p]$ is the
probability that a path from A to B is open. Repeating this process n
times, $[(z - 1)p]^n$ is the probability that a path of n steps from A is
open. If $[(z - 1)p] < 1$, then $[(z - 1)p]^n \to 0$ as $n \to \infty$. This implies
that the probability for an infinite cluster spanning the lattice is zero.
Therefore

$$p_c = \frac{1}{z - 1}. \qquad \text{III.7}$$

This result is valid for both the site and bond percolation problems.

What is the average cluster size S when p is near the threshold p_c?
To calculate S, first note that there are four branches ($z = 4$) from the
origin O leading to each site A, and introduce T the average size of the
cluster for each branch. T is the average number of sites to which the
origin is connected through a single branch A. A single branch can be
further subdivided into three other infinite branches, T again being
equal to the average size of a cluster of each branch. Leaving site A, a
neighbor site is either an unoccupied site with probability $(1 - p)$ or
occupied with probability p. The unoccupied sites contribute 0 weight to
the branch, while the occupied sites contribute a weight $(1 + 3T)$ to the

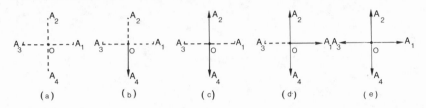

Fig. III.7. The 5 different possibilities of connections between a site O and its 4 neighbors.

branch: 1 for the site itself and $3T$ for the three branches emanating from that site. Thus

$$T = (1 - p) \cdot 0 + p \cdot (1 + 3T), \quad T = \frac{p}{1 - 3p}.$$

The size of the cluster emanating from the origin is thus 0 if the site is empty, and $(1 + 4T)$, if the origin is occupied. Thus:

$$S = (1 - p) \cdot 0 + p \cdot (1 + 4T), \quad S = p\left(\frac{1 + p}{1 - 3p}\right).$$

Because $p_c = 1/3$, it is apparent that S varies as $(p_c - p)^{-1}$ for values of p below p_c. S diverges as $p \to p_c$, with a critical exponent $x = 1$:

$$S \approx (p - p_c)^{-1}.$$

By analogous reasoning, we can compute the fraction P of sites which are attached to the infinite cluster, when the medium is above the percolation threshold p_c. P is also the probability that the original site O belongs to the infinite cluster. Consider the original site O and the four adjacent sites A (fig. III.7), and introduce Q, the probability that a path from O passing through a neighboring site A_i will be interrupted somewhere. Examination of figure 7 shows that the probability P is determined by the probability of the three configurations c, d, or e. In this figure an arrow indicates an infinite chain. It is clearly necessary that O be part of at least two infinite chains, if it is to be part of the percolating cluster.

The probability of having an arrow between O and a site A is by definition $(1 - Q)$. For the case III.7c, the probability is $6Q^2(1 - Q)^2$ because there are six combinations possible. For the case d it is $4Q(1 - Q)^3$, and for case e, $(1 - Q)^4$. Thus

$$P = 6Q^2(1 - Q)^2 + 4Q(1 - Q)^3 + (1 - Q)^4. \qquad \text{III.8}$$

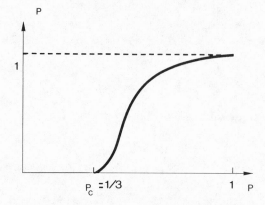

Fig. III.8. Probability of percolation P as a function of the site (or bond) occupation probability p.

The probability Q is in fact a function of p. Actually, a chain link from O through a site A is interrupted if O and A are not connected (probability $1 - p$), or if O and A are connected but the three other chain links to A are interrupted at a distance (probability pQ^3). This results in

$$Q = (1 - p) + pQ^3.$$

Equation III.8 has one simple solution, $Q = 1$, which implies $P = 0$: the system is below the percolation threshold. This equation admits two other solutions, only one of which has physical meaning

$$Q = -\frac{1}{2} + \sqrt{\frac{1}{p} - \frac{3}{4}}. \qquad \text{III.9}$$

Q decreases from 1 to 0 as p increases from $p_c = 1/3$ to 1. For $p < p_c$, $Q = 1$. The two relations $P(Q)$ and $Q(p)$ can be combined to obtain $P(p)$. Near the percolation threshold p_c, it can be shown that P varies as $(p - p_c)^2$, (fig. III.8). The critical exponent is 2:

$$p \rightarrow p_c, P \approx (p - p_c)^2. \qquad \text{III.10}$$

PERCOLATION THROUGH A FRACTURED MEDIUM

Percolation theory provides a method for describing strongly heterogeneous media such as rocks, where the effects of a threshold are often observed. Threshold effects are most apparent in the transport and elastic properties of rocks. In this section we shall take a more detailed look at the examples of permeability and elastic moduli of fractured medium.

Fractured Media

Consider an intact rock which contains fractures of various sizes. When the fractures are not too numerous, they are predominantly isolated from each other and the rock permeability is zero. As the density of

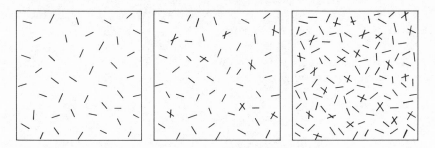

Fig. III.9. Evolution of a fractured medium when the fracture density is increased.

fractures increases, the degree of fracture interconnectivity also increases. Beyond a critical density, a fraction f of the fractures forms an "infinite cluster" (fig. III.9) and the permeability is no longer zero. If we use the preceding model (Bethe lattice with $z = 4$) to interpret the fracture network, we know that $f = P$ and $p_c = 1/3$, where p is the probability that a bond between two adjacent sites is occupied. In this example p can be interpreted as the probability that two fractures are connected. Above the critical probability p_c, the permeability is directly proportional to the fraction P of connected fractures. These considerations are also applicable to the electrical conductivity of the fracture network, if the conductivity is due to fluid in the fractures.

For simplicity let us describe a fractured medium by assuming that all fractures are disk shaped, of radius c, aperture $2w$, and of number density N (proportional to $1/l^3$ where l is the average spacing between fractures) (fig. III.10). The fractures are such that $w \ll c$. How can we

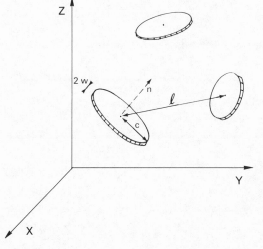

Fig. III.10. Descrpition of a fractured medium: the elementary fractures are discs of radius c and aperture $2w$.

express p as a function of c, w, and N? It is clear that p increases with N and with c: the more fractures there are and the larger their radii, the greater the probability of their intersection. With N and c it is possible to construct only one dimensionless quantity, Nc^3. Because $p = 0$ when N or c are zero, one concludes that p varies as Nc^3:

$$p \approx Nc^3 \equiv \frac{c^3}{l^3}.$$

Domain of Percolation and Permeability

A percolation threshold occurs when Nc^3 reaches a critical value ($1/3$ in the case of a Bethe network with $z = 4$). It is possible to determine the threshold conditions exactly by computing the probability p. The computation of p requires introducing the notion of excluded volume. For example, consider a disk-shaped fracture that is centered at O and has a radius c. What is the largest volume around O, such that if the center O' of a second fracture (of radius c and random orientation) lies within this volume, these two fractures will intersect? This is by definition the excluded volume for a disk V_{ex} (De Gennes, 1976):

$$V_{ex} = \pi^2 c^3. \qquad \text{III.11}$$

For a fracture density N the average number of intersections for one fracture is $v = NV_{ex}$. The Bethe network indicates that the probability for a fracture to be isolated is $(1 - p)^4$. This probability can also be expressed in terms of v. Let V_o be a very large volume, in the interior of which there is a random distribution of points of density N (the centers of the disks or fractures). The probability that a random point in V_o finds itself in the interior of a smaller volume V, which is included in V_o, is V/V_o. If we consider n random points in V_o, the probability for m of them being in V is $p_m = C_n^m \left(\dfrac{V}{V_o}\right)^m \left(1 - \dfrac{V}{V_o}\right)^{n-m}$ where $C_n^m = \dfrac{n!}{(n-m)!m!}$. In taking the limit of this expression as $n \to \infty$, $V_o \to \infty$, with $n/V_o = N$, one obtains:

$$\lim p_m = \frac{1}{m!}(NV)^m \left(1 - \frac{NV}{n}\right)^n = \frac{1}{m!}v^m e^{-v}.$$

For $m = 0$, $p_o =$ the probability of a fracture being isolated $= e^{-v} = (1 - p)^4$. Therefore

$$p = 1 - e^{-v/4}. \qquad \text{III.12}$$

When $v \ll 1$, $p \approx v/4$. Because $v = NV_{ex} = N\pi^2 c^3$, equation III.12 is the desired relation between p, c, and N. Utilizing $N \approx 1/l^3$ and

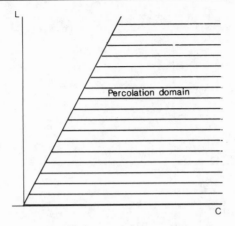

Fig. III.11. Percolation domain in the plane (l, c): there is no percolation if l is too large (small fracture density) or if c is too small (very small fractures.

$p \approx \nu/4$, the condition for the percolation threshold $p_c = 1/3$ divides the "plane" (l, c) into two domains: the domain of percolation corresponds to values (fig. III.11):

$$\frac{c}{l} \geq \left(\frac{4}{3\pi^2} \right)^{1/3}$$

and

$$p \cong \frac{\pi^2}{4} \frac{c^3}{l^3}. \qquad \text{III.13}$$

According to equation III.10, when p is above but close to p_c, the permeability is proportional to $(p - p_c)^2$. This is also true for the electrical conductivity, if the conductivity is due to fluid in the fractures.

Elastic Moduli

Above the percolation threshold, the elastic moduli M vary with the percolation parameter p according to the power law

$$M \approx (p - p_c)^n.$$

This expression is analogous to that for the electrical conductivity and permeability. However, the exponent n is not the same as the exponent for conductivity. In three dimensions, the conductivity exponent is 2, while in two dimensions it is 1.3. Simple experiments carried out on slabs of resins pierced by holes permit the determination of the elastic modulus exponent in two dimensions (fig. III.12). The value of the

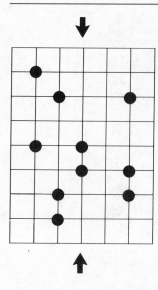

Fig. III.12. Resin plate covered with a network: the holes are punched with probability p. M varies with p.

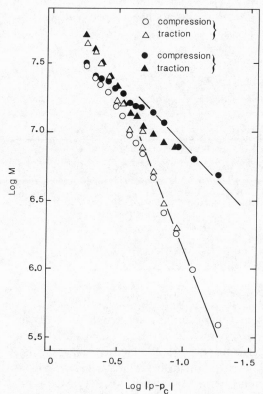

Fig. III.13. Law of variation $M \sim (p - p_c)^n$. The empty symbols correspond to measurements with the holes empty. The filled symbols are measurements where the cylinders were reintroduced into the holes.

exponent is found to be 2.1 when the holes are empty, a value that is significantly different from the value 1.3 observed for the conductivity (fig. III.13).

To simulate a fractured medium, which necessarily has a very small porosity, the extracted cylinders can be reintroduced back into the holes. In this case the exponent is found to be very close to the conductivity exponent ($n = 0.9$) (Chelidze et al., 1988).

Percolation effects are very different in elasticity because the constitutive relation between constraints and deformations has a different tensorial nature, as compared to the relation between the flux and electric field in conductivity. Additional degrees of freedom due to rotations leads one to talk about "vector percolation" in the case of elasticity, while conductivity is concerned with "scalar percolation."

PROBLEMS

III.a. Effective Medium Direct Calculation

Calculate the term $\dfrac{\bar{\varepsilon}^{(2)}}{\bar{\varepsilon}}$ for spherical pores imbedded in an isotropic medium at a low porosity. For the solution of this problem use the results of chapter IV. Consider a spherical pore (radius R_2) at the center of a larger sphere of radius R_1. Assume that a hydrostatic compression $-\Delta p$ is applied to the surface of the larger sphere. Utilize the equation of equilibrium appropriate for this geometry, $\vec{\nabla}(\vec{\nabla} \cdot \vec{u}) = 0$.

(1) Show that the radial displacement is given by $u = ar + \dfrac{b}{r^2}$.

(2) Calculate a and b utilizing the boundary conditions at $r = R_1$ and $r = R_2$: $\sigma_{rr}(R_2) = 0$ and $\sigma_{rr}(R_1) = -\Delta p$.

(3) Derive the result $\dfrac{\bar{\varepsilon}^{(2)}}{\bar{\varepsilon}} = \left(1 + \dfrac{3K_1}{4\mu_1}\right)$.

III.b. Geometric, Arithmetic, and Harmonic Averages

For the case of a two-component system with proportions v_1 and v_2, one can define the average property M by the relation $M^x = v_1 M_1^x + v_2 M_2^x$ with $-1 \le x \le 1$. Let $v_1 = \phi$ and $v_2 = 1 - \phi$.

(1) Show that the cases $x = 0$, 1, and -1 correspond to the geometric, arithmetic, and harmonic averages.

(2) Graph the variation of M as ϕ varies from 0 to 1 for the range of values of x defined previously.

III.c. Excluded Volume for Disks

Consider a disk-shaped fracture of radius c. The excluded volume is the volume V_{ex} such that if a second fracture is introduced into the interior of this volume, the two will be in contact. Consider a second fracture also of radius c. Let β be the angle between the fractured planes and z be the height of the center of the second fracture above the plane of the first (normal distance from the plane of the first fracture).

(1) Show that if z is constant, the center of the second fracture must be in the interior of a surface given by

$$S = (4r'c + \pi c^2) \quad \text{where} \quad r' = \left[c^2 - \left(\frac{z}{\sin \beta} \right)^2 \right]^{1/2}.$$

(2) Calculate the two limits z_o and $-z_o$ for the allowable values of z.
(3) If all angles β are equally probable, show that $V_{ex} = \pi^2 c^3$.

REFERENCES

Beran, M. J., and J. J. McCoy. 1970. Mean field variations in a statistical sample of heterogeneous linearly elastic solids. *Int. J. Solids and Structures* 6:1035–54.

Chelidze, T., T. Reuschle, M. Darot, and Y. Guéguen. 1988. On the elastic properties of depleted refilled solids near percolation. *J. Physics C.: Solid State Phys.* 21:L1007–L1010.

De Gennes, P. G. 1976. *The Physics of Liquid Crystals*. Oxford University Press.

Hashin, Z. 1983. Analysis of Composite Materials, A Survey. *Journal of Applied Mechanics* 50:481–505.

Landau, L. D., and E. M. Lifshitz. 1967. *Theory of Elasticity*. Moscow: Editions Mir.

Mandelbrot, B. B. 1982. *The Fractal Geometry of Nature*. San Francisco: Freeman.

Stauffer, D. 1985. *Introduction to Percolation Theory*. London: Taylor and Francis.

IV. Mechanical Behavior of
Dry Rocks

THE MECHANICAL behavior of rocks depends crucially on the fact that they are a heterogeneous and porous media. When submitted to stresses, rocks deform in a complicated manner. In general it is not possible to describe all of the deformations that occur with the same laws. Strain analysis must take into account scale effects, both in time and space. Thus it is essential to specify the validity field for each particular mechanical response. The same rock may show an elastic response when submitted to a rapidly varying stress (about one-second period) and a plastic response when submitted to slowly varying stresses (about one-million-year period, i.e., $3 \cdot 10^{13}$ seconds). This wide range of scales (both in time and space) is typical of geophysics and geology, and implies that much care is required when considering deformations over long time scales.

In this chapter, we consider only dry rocks. The important effects due to fluids will be examined in chapter VI.

STRESS AND STRAIN

To investigate the mechanical behavior of a given medium, the relationship between stresses and strains must be examined. We begin by defining these two concepts.

Stress

The concept of stress is the generalization of the one-dimensional concept of tension in a string (fig. IV.1). Consider a solid with an external surface which is submitted to a distribution of forces. Imagine that this solid is cut in two parts. To maintain each of these parts in equilibrium, it is necessary to apply a distribution of forces $d\vec{F} = \vec{\sigma} \, dS$ on the cut surfaces. Figure IV.2b shows how the left part of the solid is in equilibrium under the initial distribution of forces (on its external surface) and the new distribution $d\vec{F} = \vec{\sigma} \, dS$ (on the new surface S). The sum of all elementary forces $d\vec{F}$ is $\vec{F}$, the analogue of tension in a string.

Because of the principle of action and reaction, the second half of the solid (the right part) is in equilibrium if an equal but opposite distribution of forces is applied on (S). The vector $\vec{\sigma}$ is the stress vector (or traction). In general $\vec{\sigma}$ varies from one point to another: $\vec{\sigma} = \vec{\sigma}(x, y, z)$.

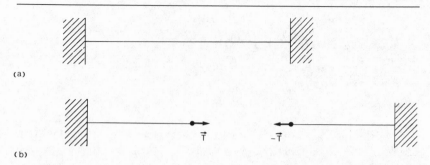

Fig. IV.1. Concept of tension in a rope: (*a*) intact rope; (*b*) cut rope. In order to maintain both segments in equilibrium, forces $+\vec{T}$ and $-\vec{T}$ must be applied at their extremities.

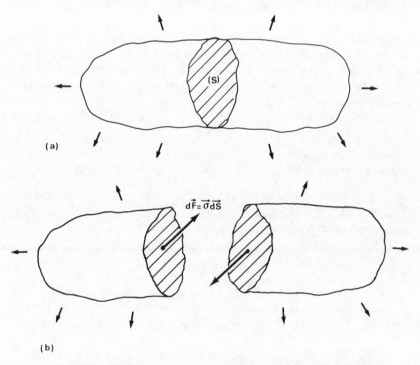

Fig. IV.2. Solid with forces applied on its external surface. (*a*) S is an arbitrary surface within the solid. (*b*) The solid has been cut along S. In order to maintain equilibrium, a distribution of forces $d\vec{F} = \vec{\sigma}d\vec{S}$ must be applied on S.

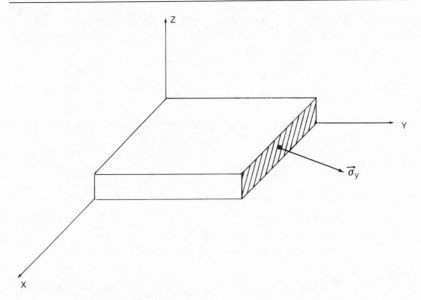

Fig. IV.3. $\vec{\sigma}_y$ is the stress vector or traction on the infinitesimal area $dS = dx\,dy$. The normal to this surface is parallel to the Oy axis.

If we consider a small parallelopiped in a reference frame $Oxyz$, then three different stress vectors have to be introduced, each of them being relative to one face perpendicular to one axis. For instance, $\vec{\sigma}_y$ is shown in figure IV.3. This stress vector is relative to the face perpendicular to the Oy axis. Note that $\vec{\sigma}_y$ is not in general parallel to the Oy axis. It has three components: $\sigma_{yx}, \sigma_{yy}, \sigma_{yz}$. The same is true for $\vec{\sigma}_x$ and $\vec{\sigma}_z$. Components such as σ_{xx} are called normal components; components such as σ_{xy} are called shear components.

Consider the infinitesimal parallelopiped shown on figure IV.3. At equilibrium, the moments of rotation must sum to zero. For example, taking moments with respect to axis Oz implies

$$(\sigma_{xy} - \sigma_{yx})\, dx\,dy\,dz = 0 \Rightarrow \sigma_{xy} = \sigma_{yx}.$$

By considering moments with respect to axes Ox and Oy one arrives at the general result

$$\sigma_{ij} = \sigma_{ji} \qquad\qquad \text{IV.1}$$

where i and j stand for x, y, or z. There are six independent quantities σ_{ij} which are the components of a second-order symmetric tensor called the stress tensor.

The equilibrium of the parallelopiped also demands that the resultant forces vanish. If only forces in the x direction are considered, equilibrium requires

$$\frac{\partial \sigma_{xx}}{\partial x} + \frac{\partial \sigma_{xy}}{\partial y} + \frac{\partial \sigma_{xz}}{\partial z} = 0$$

when no body forces are present. The above equation arises from the fact that stresses on opposite faces perpendicular to the x-axis differ by the amount $\dfrac{\partial \sigma_{xx}}{\partial x} dx$. Similar differences exist for faces perpendicular to the y and z axes. If a distribution of body forces $\vec{F}$ also exists, the equation for equilibrium becomes:

$$\frac{\partial \sigma_{ij}}{\partial x_j} + F_i = 0. \qquad\qquad \text{IV.2}$$

Perhaps the most common example of a body force is gravity. Then $\vec{F} = \rho \vec{g}$, where ρ is the density and $\vec{g}$ the acceleration due to gravity. In this case, the force on an infinitesimal volume $dv = dx\,dy\,dz$ would be $\vec{F}\,dv = \rho \vec{g}\,dv$.

By analyzing the equilibrium of a tetrahedron (fig. IV.4), the stress vector $\vec{\sigma}$ on a plane defined by its normal vector $\vec{n}$ is found to be

$$\sigma_i = \sigma_{ij} n_j. \qquad\qquad \text{IV.3}$$

For example, the sum of forces along the x-axis can be written as:

$$\sigma_x \, dS = \sigma_{xj} \, dS_j.$$

The contribution of body forces is zero in the limit $dS \to 0$. The area of ABC is dS and dS_x, dS_y, dS_z are the areas of OBC, OAC, OAB, respectively. Because $\vec{n}$ is the normal to ABC, $dS_x = n_x\,dS =$ the projection of dS on the plane perpendicular to Ox, with similar results for dS_y and dS_z.

Stress is a force per unit area. Thus the concept of stress is an extension of the concept of pressure. When $\sigma_{xx} = \sigma_{yy} = \sigma_{zz}$ and simultaneously all other σ_{ij} are zero, we can write:

$$\sigma_{ij} = -p\, \delta_{ij}$$

where p is the hydrostatic pressure. We shall use the convention that stresses are negative when they refer to a compressive state and positive for extension. In the earth's crust, "lithostatic" equilibrium is defined by

$$\sigma_{xx} = \sigma_{yy} = \sigma_{zz} = \rho g z.$$

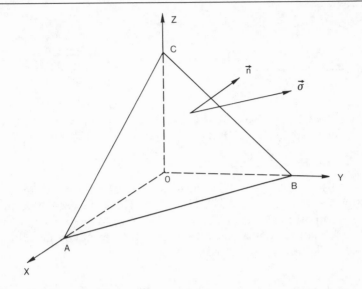

Fig. IV.4. Infinitesimal tetrahedron defined by three orthogonal planes and a plane normal to $\vec{n}$. $\vec{\sigma}$ is the stress vector relative to the plane normal to $\vec{n}$. The area of ABC is dS.

There are no other non-zero components. In this particular stress state only the weight of the overlying rocks comes into play. Because the density ρ of rocks is high, σ_{zz} increases significantly with depth. At a depth of 10 km:

$$\sigma_{zz} = 2500 \times 9.81 \times 10^4 = 22 \times 10^7 \text{ Pa or 220 MPa}$$

for a density $\rho = 2500$ kg m^{-3}.

The S.I. unit for pressure is Pascal (1 Pa = 1 Newton m^{-2}). Another commonly used pressure unit is the bar (1 bar = 10^6 dynes cm^{-2} = 10^5 Pa). In bars, the lithostatic pressure at 10 km is 2.2 kbars.

DEFORMATIONS

A deformed solid is one which has experienced a change of shape. When stresses are applied, different points of a solid move with respect to each other and their separations change. Angles between straight lines, such as AB and AC in the undeformed solid (fig. IV.5), are also modified. These changes in lengths and angles are the basic parameters which allow us to calculate how $ds^2 = (AB)^2 = dx^2 + dy^2 + dz^2$ transforms into $(A'B')^2$ after deformation. Point A is transformed into A'

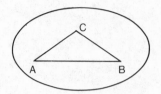 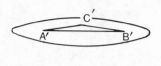

Fig. IV.5. Undeformed and deformed solid. Deformation is expressed through a change in length ($AB \neq A'B'$) and a change in angle. (ACB angle $\neq A'C'B'$ angle.)

(i.e., $\overrightarrow{AA'} = \vec{u}$) and point B, which is infinitesimally close to A, is transformed into B' (i.e., $\overrightarrow{BB'} = \vec{u} + \overrightarrow{du}$). The displacement vector $\vec{u}$ varies with position (x, y, z): $\vec{u} = \vec{u}(x, y, z)$. We assume however that $\vec{u}$ is always small as compared to the size of the object considered (rock sample, rock mass, etc.). We want to calculate

$$(A'B')^2 = (dx + du_x)^2 + (dy + du_y)^2 + (dz + du_z)^2.$$

Noting that $du_x = \dfrac{\partial u_x}{\partial x} dx + \dfrac{\partial u_x}{\partial y} dy + \dfrac{\partial u_x}{\partial z} dz$ (and using similar expressions for du_y and du_z), one derives:

$$(A'B')^2 = dx^2 + dy^2 + dz^2 + 2\frac{\partial u_x}{\partial x} dx^2 + 2\frac{\partial u_y}{\partial y} dy^2 + 2\frac{\partial u_z}{\partial z} dz^2$$

$$+ 2\left(\frac{\partial u_x}{\partial y} + \frac{\partial u_y}{\partial x}\right) dx\,dy + 2\left(\frac{\partial u_x}{\partial z} + \frac{\partial u_z}{\partial x}\right) dx\,dz$$

$$+ 2\left(\frac{\partial u_z}{\partial y} + \frac{\partial u_y}{\partial z}\right) dy\,dz + \text{negligible terms.}$$

The neglected quantities are of the form $\dfrac{\partial u_x}{\partial x} \dfrac{\partial u_x}{\partial y} dx\,dy$ and contain the product of two derivatives $\dfrac{\partial u_i}{\partial x_j}$ and $\dfrac{\partial u_i}{\partial x_k}$. Because we have assumed that $\vec{u}$ is a small vector, we can consider that $\left|\dfrac{\partial u_i}{\partial x_j}\right| \ll 1$, so that partial derivatives are first-order quantities and the neglected terms are of second order. With these assumptions, one finds the following expres-

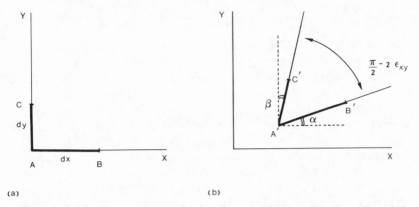

Fig. IV.6. (*a*) Undeformed solid: AB and AC are orthogonal and have lengths dx and dy. (*b*) deformed solid: AB is transformed into $A'B'$ and AC into $A'C'$. The quantity ε_{xy} measures the variation of the right angle xOy. Quantities ε_{xx} and ε_{yy} represent the changes in length, per unit initial length, of vectors parallel to the Ox and Oy axes.

sion for the change in separation between two adjacent points in the medium:

$$(A'B')^2 - (AB)^2 = \left(\frac{\partial u_i}{\partial x_j} + \frac{\partial u_j}{\partial x_i} \right) dx_i \, dx_j = 2\varepsilon_{ij} \, dx_i \, dx_j.$$

There is an implied summation over the repeated indices i and j. The infinitesimal deformation tensor ε_{ij} is defined as

$$\varepsilon_{ij} = \frac{1}{2} \left(\frac{\partial u_i}{\partial x_j} + \frac{\partial u_j}{\partial x_i} \right).$$ IV.4

If we examine two infinitesimal line elements dx and dy which are orthogonal in the undeformed solid, we see that terms of the type ε_{xx} describe the relative change in length and ε_{xy} the change in angle (fig. IV.6):

$$\frac{A'B' - AB}{AB} = \varepsilon_{xx} \text{ and } \alpha = \frac{\partial u_y}{\partial x}, \beta = \frac{\partial u_x}{\partial y} \text{ where } \varepsilon_{xy} = \frac{\alpha + \beta}{2}.$$

It follows that $\varepsilon_{kk} = \varepsilon_{xx} + \varepsilon_{yy} + \varepsilon_{zz}$ represents the relative change in volume $\dfrac{\Delta v}{v}$. For a parallelopiped of sides l_1, l_2, l_3: $v = l_1 l_2 l_3$ and thus

$$\frac{\Delta v}{v} = \frac{\Delta l_1}{l_1} + \frac{\Delta l_2}{l_2} + \frac{\Delta l_3}{l_3} = \varepsilon_{xx} + \varepsilon_{yy} + \varepsilon_{zz}.$$

Tensors

Stresses and deformations are defined as second-order, symmetric tensors. A tensor is a mathematical object which is characterized by a number of indices (tensor rank or order) and which obeys well-defined rules for how the tensor components change when there is a transformation of coordinates. Tensors can represent physical quantities. For example, scalars like mass are zero-order tensors (they are invariant when the coordinate system changes). Vectors, such as forces, are first-order tensors (their components change when there is a change in coordinate system, i.e., a rotation). Tensors σ_{ij} and ε_{ij} are second-order tensors. When there are two coordinate systems specified by the unit vectors $(\vec{e}_1, \vec{e}_2, \vec{e}_3)$ and $(\vec{e}_1'', \vec{e}_2'', \vec{e}_3'')$, the stress tensor σ_{kl}' in the second coordinate system can be expressed in terms of the stress tensor σ_{kl} in the first system as:

$$\sigma_{ij}' = \alpha_{ik}\alpha_{jl}\sigma_{kl}$$

where (α_{mn}) is the matrix such that $\vec{e}_m'' = \alpha_{mn}\vec{e}_n$. There always exists a coordinate system where σ_{ij} and ε_{ij} are diagonal tensors. This is the principal directions coordinate system.

For more details, the reader can consult fundamental textbooks such as Landau and Lifchitz, *Elastic Theory* (1967) and Fung, *Foundations of Solid Mechanics* (1965).

ELASTIC BEHAVIOR

For short time scales (less than 10^5 sec), the mechanical behavior of rocks is described quite well by the theory of elasticity. This is only true if the temperature, pressure, and stress conditions are "moderate." We shall later define more precisely what is meant by "moderate." In addition, the deformations must be very small. When deformations reach a threshold of 10^{-2}, and in many cases less than this value, the behavior is no longer elastic. An elastic wave of extremely small amplitude offers an example of almost ideal elastic behavior.

Shear and Bulk Moduli

It is useful to consider two basic elementary stress conditions: (1) hydrostatic pressure described by $\sigma_{ij} = -p\delta_{ij}$, and (2) shear stress with only σ_{xy}(or σ_{yz}, or σ_{xz}) not equal to zero. Then any more general stress state can be reduced to a linear combination of these two simpler cases. In the first case, the deformation is simply a change in volume (described by ε_{kk} as shown above) and in the second, there is a change in angle (described by the off diagonal ε_{ij}). The basic assumption of elastic

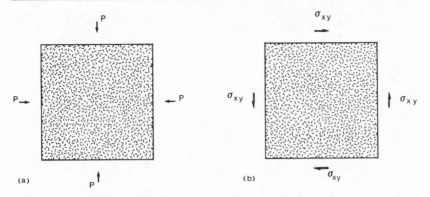

Fig. IV.7. (*a*) Hydrostatic compression. (*b*) Pure shear.

theory is that deformations are proportional to stresses and that they are reversible (i.e., if stresses are suppressed, deformations vanish). These assumptions are embodied in Hooke's law:

$$\varepsilon = \frac{\sigma}{M}. \qquad\qquad \text{IV.5}$$

The parameter M is an elastic modulus, characteristic of the solid being considered. Because ε is dimensionless, M has the units of pressure (i.e., one Pa or one bar). For the two elementary stress states, there are two distinct moduli (fig. IV.7):

(*a*) a hydrostatic compression: $\varepsilon_{kk} = \dfrac{\sigma_{kk}}{3K}$, when $M = 3K$

(*b*) a shear deformation (with $i \neq j$): $\varepsilon_{ij} = \dfrac{\sigma_{ij}}{2\mu}$, when $M = 2\mu$. $\qquad$ IV.6

Note that $\left(-\dfrac{\sigma_{kk}}{3}\right)$ is the average "pressure" and that $K = \chi^{-1}$, where $\chi = -\dfrac{1}{V}\dfrac{\partial V}{\partial P}$ is the coefficient of compressibility. The compressibility is a measure of the fractional volume change $\dfrac{\delta V}{V}$ when the pressure is increased by δP. Modulus K is known as the bulk modulus, and modulus μ is known as the shear modulus.

Elastic moduli are measured using either traction-compression experiments (static deformation) or elastic pulse propagation (dynamic deformation). Static moduli of rocks are in general lower than dynamic moduli. As we shall see later on, this is due to the fact that rocks always contain cracks. Static moduli determined in traction differ from those determined in compression for the same reason. Bulk modulus K is

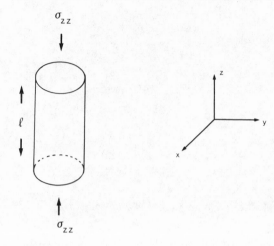

Fig. IV.8. Uniaxial compression experiment: measurement of Young's modulus $E = \sigma_{zz}/\varepsilon_{zz}$, with $\varepsilon_{zz} = \delta l/l$. Poisson's ratio is defined as $\nu = -\varepsilon_{yy}/\varepsilon_{zz}$.

obtained through isotropic compression experiments. Uniaxial compression is however simpler to perform, so that most frequently it is the Young's modulus E which is measured in the static regime (fig. IV.8).

By definition, $\varepsilon_{zz} = \dfrac{\sigma_{zz}}{E}$ for the experimental conditions shown in figure IV.8. In the same uniaxial compression experiment, one also defines the Poisson ratio ν through $\nu = -\dfrac{\varepsilon_{yy}}{\varepsilon_{zz}}$ or equivalently, $\varepsilon_{yy} = -\nu\dfrac{\sigma_{zz}}{E}$. Young's modulus E and Poisson ratio ν can be expressed in terms of K and μ. By summing the deformations due to the uniaxial compressions $\sigma_{xx}, \sigma_{yy}, \sigma_{zz}$, one gets the total deformation $\varepsilon_{zz} = \dfrac{\sigma_{zz}}{E} - \nu\dfrac{\sigma_{xx}}{E} - \nu\dfrac{\sigma_{yy}}{E}$. Similar results can be obtained for ϵ_{xx} and ε_{yy}.

$$E\varepsilon_{xx} = \sigma_{xx} - \nu(\sigma_{yy} + \sigma_{zz}). \qquad \text{IV.7}$$

Summing the diagonal terms leads to: $E\varepsilon_{kk} = \sigma_{kk}(1 - 2\nu)$. Comparing this expression with the definition of K, one arrives at the relation $3K = \dfrac{E}{1 - 2\nu}$.

It is also possible to express μ as a function of E and ν. To do so, consider the simple two-dimensional example described in figure IV.9. Here the only non-zero stress is $\sigma_{xy} = \sigma$, hence $\varepsilon_{xy} = \varepsilon = \sigma/2\mu$ is the only non-zero deformation. Both tensors, stress and strain, are diagonal in the $Ox'y'$ coordinate system (fig. IV.9b). In this principal direction

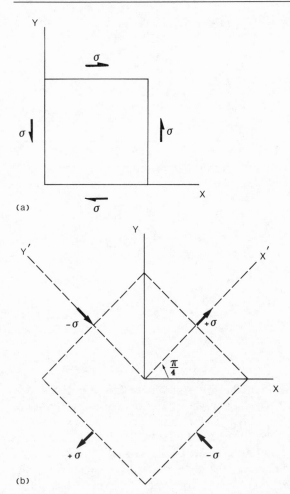

Fig. IV.9. (a) Two-dimensional stress field for the case where $\sigma_{xy} = \sigma$ and $\sigma_{xx} = \sigma_{yy} = 0$. (b) Principal directions of the above stress field.

coordinate system, the stresses are $\pm\sigma$ and the strains are $\pm\varepsilon$. There is an extension in the x' direction and a compression in the y' direction. Thus $E\varepsilon_{x'x'} = \sigma_{x'x'} - \nu\sigma_{y'y'}$ with $\sigma_{x'x'} = +\sigma$, $\sigma_{y'y'} = -\sigma$, and $\varepsilon_{x'x'} = \varepsilon$. Combining we find $\varepsilon = \dfrac{\sigma}{E}(1 + \nu)$. Comparing this with $\varepsilon = \dfrac{\sigma}{2\mu}$, one derives $2\mu = \dfrac{E}{1 + \nu}$.

Inverting the above relations between K, μ and E, ν, one finds $\nu = \dfrac{3K - 2\mu}{2(3K + \mu)}$. Tables IV.1 and IV.2 summarize these definitions and relations between the four elastic constants.

Table IV.1. DEFINITIONS OF ELASTIC CONSTANTS

	K	μ	E	ν
Type of Constraint	Isotropic Compression δP	Shear σ_{xy}	Uniaxial Compression σ_{zz}	Uniaxial Compression σ_{zz}
Definition	$-V\dfrac{\delta P}{\delta V}$	$\dfrac{\sigma_{xy}}{2\varepsilon_{xy}}$	$\dfrac{\sigma_{zz}}{\varepsilon_{zz}}$	$-\dfrac{\varepsilon_{xx}}{\varepsilon_{zz}}$

Table IV.2. RELATIONS BETWEEN K, μ AND E, ν

	K, μ	E, ν
K	K	$\dfrac{E}{3(1-2\nu)}$
μ	μ	$\dfrac{E}{2(1+\nu)}$
E	$\dfrac{9K\mu}{3K+\mu}$	E
ν	$\dfrac{3K-2\mu}{2(3K+\mu)}$	ν

Typical values of these coefficients, for three different rocks, are given in table IV.3. The Poisson coefficient is a dimensionless parameter which varies between 0 and $1/2$ (the limit $\nu \to 1/2$ corresponds to that of a fluid when μ and $E \to 0$). On the average, elastic moduli vary in the range $10^{10}-10^{11}$ Pa (100 kb–1 Mb).

Dynamic measurements utilize the ultrasonic wave propagation technique. Consider a compression wave traveling along the x-axis. The displacement vector is $u_x(x,t) = e^{i(kx-\omega t)}$ for a plane wave with phase velocity $c = \omega/k$. The associated strain is $\varepsilon_{xx} = \dfrac{\partial u_x}{\partial x}$. All the other ε_{ij}

Table IV.3. TYPICAL VALUES OF ELASTIC CONSTANTS

	ν	E $(10^{10}Pa)$	μ $(10^{10}Pa)$	K $(10^{10}Pa)$
Basalt (Pigash)	0.19	6.24	2.38	3.35
Granite (Barre)	0.10	3.04	1.38	1.26
Sandstone (Cherokee)	0.10	3.99	1.81	1.66

Source: CRC Handbook of Physical Properties of Rocks, vol. 2, 1982, R. S. Carmichael ed. Boca Raton, Fla.: CRC Press.

are zero. Using IV.2 and assuming zero body forces, but a non-zero acceleration, one can derive the dynamic equilibrium equation:

$$\frac{\partial \sigma_{xx}}{\partial x} = \rho \frac{\partial^2 u_x}{\partial t^2} \quad \text{with} \quad E\varepsilon_{xx} = \sigma_{xx} - \nu(\sigma_{yy} + \sigma_{zz}).$$

Because $\varepsilon_{yy} = \varepsilon_{zz} = 0$ and by symmetry $\sigma_{yy} = \sigma_{zz}$, IV.7 implies:

$$\sigma_{yy} = \sigma_{zz} = \frac{\nu}{1 - \nu}\sigma_{xx}.$$

With the above relation:

$$E\varepsilon_{xx} = \sigma_{xx} - \frac{2\nu^2}{1 - \nu}\sigma_{xx} = \frac{(1 + \nu)(1 - 2\nu)}{1 - \nu}\sigma_{xx}.$$

Thus the dynamic equilibrium equation becomes:

$$\left[\frac{E(1 - \nu)}{(1 + \nu)(1 - 2\nu)} \right] \frac{\partial^2 u_x}{\partial x^2} = \rho \frac{\partial^2 u_x}{\partial t^2}.$$

This is the propagation equation for a compression wave (P-wave) in the x-direction. Because $u_x(x,t) = e^{i(kx - \omega t)}$, one obtains for the P wave velocity, $V_p = \frac{\omega}{k} = \left[\frac{E(1 - \nu)}{(1 + \nu)(1 - 2\nu)\rho} \right]^{1/2}$. Substituting K and μ for E and ν, V_p becomes:

$$V_p = \sqrt{\frac{K + \frac{4}{3}\mu}{\rho}}. \qquad\qquad \text{IV.8}$$

Let us now consider a shear deformation propagating along the x-axis. If $u_y(x,t) = e^{i(kx - \omega t)}$ is the displacement, then

$$\varepsilon_{xy} = \frac{1}{2}\frac{\partial u_y}{\partial x}, \quad \sigma_{xy} = \mu \frac{\partial u_y}{\partial x}, \quad \frac{\partial \sigma_{xy}}{\partial x} = \rho \frac{\partial^2 u_y}{\partial t^2}.$$

The propagation equation for the shear wave (S-wave) becomes

$$\mu \frac{\partial^2 u_y}{\partial x^2} = \rho \frac{\partial^2 u_y}{\partial t^2}$$

and the S-wave velocity is V_s:

$$V_s = \sqrt{\frac{\mu}{\rho}}. \qquad\qquad \text{IV.9}$$

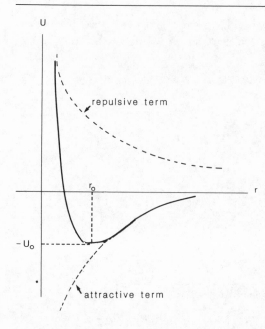

Fig. IV.10. Plot of U the potential energy per molecule of solid versus r, the interatomic distance. At equilibrium $r = r_o$, $U = -U_o$.

By measuring the propagation time for a pulse to traverse a sample, either in compression or in shear, one can measure K and μ, if ρ is known. The wavelength must be large compared to the grain size and the pore or crack sizes. Given that elastic moduli vary in the range 10^{10}–10^{11} Pa and the densities in the range 2–3×10^3 kg m^{-3}, V_p and V_s are close to 5×10^3 m s^{-1}. At 1 MHz frequency, the wavelength is about 5 mm. Thus the requirement that the grain sizes and pores or cracks are smaller than 5 mm is satisfied.

Microscopic Interpretation of Intrinsic Elastic Moduli

Elastic moduli M have a precise physical meaning, which becomes clearer when the deformations are examined at the atomic scale (for a non-porous rock). In this case the elastic moduli are called the "intrinsic" moduli. Let us assume that the solid is made up of only one kind of atom (if this is not true, one can consider that a complex molecule of n atoms can be described by an equivalent atom of atomic mass equal to the average atomic mass: $\overline{m} = \dfrac{m}{n}$ where m is the mass of the real molecule). The potential energy U for such a molecule can be plotted as a function of r, the interatomic distance (fig. IV.10). The energy $U(r)$ is the sum of two energies: one being the energy of attraction due to long-range coulomb interactions, and the other being a short-range repulsive energy due to the nucleus and inner electrons.

The plot exhibits a minimum at $r = r_o$, equilibrium interatomic distance. Elastic moduli are proportional to the second derivatives of $U(r)$.

Let V be the volume of our molecule and assume $V = nr^3$. Then because $\chi = \dfrac{1}{K} = -\dfrac{1}{V}\dfrac{\partial V}{\partial P}$ and $P = -\dfrac{\partial U}{\partial V}$, one finds $K = V\dfrac{\partial^2 U}{\partial V^2}$.

Using $V = nr^3$, $P = -\dfrac{dU}{dr}\dfrac{dr}{dV} = -\dfrac{1}{3nr^2}\dfrac{dU}{dr}$. Thus the incompressibility becomes

$$K = -nr^3 \frac{dP}{dr}\frac{dr}{dV} = \frac{r}{9n}\frac{d}{dr}\left[\frac{1}{r^2}\frac{dU}{dr}\right].$$

At the equilibrium position $r = r_o$, $\dfrac{dU}{dr} = 0$. Thus

$$K_o = \frac{1}{9nr_o}\left(\frac{d^2 U}{dr^2}\right)_o. \qquad\qquad \text{IV.10}$$

Equation IV.10 shows that K_o is proportional to the second derivative of $U(r)$. A simplified calculation verifies that IV.10 gives the right order of magnitude for K_o. Approximating $\left(\dfrac{d^2 U}{dr^2}\right)_o$ by U_o/r_o^2 and taking $r_o = 2 \times 10^{-10}$ m and $(U_o/n) = 5$ eV (1 $eV = 1.6 \times 10^{-19}$ Joules), we then find $K_o \approx 10^{10}$ Pa.

A basic assumption of elastic theory is that elastic deformations are small and reversible. Such deformations result from very small atomic displacements $dr \ll r_o$. In that case, $U(r)$ can be represented near $r = r_o$ by a parabolic law

$$U = -U_o + \frac{1}{2}k(r - r_o)^2.$$

Using this approximation, the problem becomes that of a classical harmonic oscillator. Atoms are linked to each other by springs of stiffness $k = \left(\dfrac{d^2 U}{dr^2}\right)_o$. This simplified microscopic model allows us to calculate the stiffness in the elastic range. When dr becomes too large, the above assumptions are no longer valid.

Extending this model somewhat further it is possible to derive additional results. Assume the following specific form for the interatomic potential $U(r)$

$$U(r) = -\frac{\alpha}{4\pi\varepsilon_o}z_1 z_2 \frac{e^2}{r} + \frac{B}{r^p}.$$

Here α is the Magdelung constant, ε_o the dielectric permittivity of a vacuum, $z_1 e$ and $z_2 e$ the ionic charges (e is the electron charge). Substituting this relation into equation IV.10 allows K_o to be expressed as a function of the microscopic parameters.

$$K_o = -\frac{\alpha z_1 z_2 e^2 (p-1)}{36 n \pi \varepsilon_o} \left(\frac{\rho}{\overline{m}}\right)^{4/3}. \qquad \text{IV.11}$$

The solid density is taken to be $\rho = \overline{m}/r_o^3$ in IV.11. This last result is of great importance because it states that, at constant $\overline{m}$, K_o is mainly a function of ρ. There is a strong correlation between elastic moduli and density (or velocities and density) for minerals and rocks of similar average atomic mass $\overline{m}$ (see chap. VII).

Elastic Moduli of a Porous Medium

When rocks contain pores and cracks, the elastic moduli are no longer equal to the "intrinsic" moduli of the nonporous rock. In chapter III we presented some of the methods which can be used to calculate "extrinsic" moduli, or effective moduli, as a function of porosity ϕ. With the assumption that the pores are spherical, equation III.3 yields

$$K^* = K_1(1 - \beta\phi) \text{ where } \beta \text{ is a constant.}$$

The rock is considered as a composite of two phases, where the pores (the second phase) are imbedded into the main phase (1) of intrinsic modulus K_1.

Similar results are obtained for rocks containing ellipsoidal cavities. Walsh (1965) was the first to derive the expression for K^* for that case, but later derivations have also been given by several others (Henyey and Pomphrey, 1982). The general result has the form

$$K^* = K_1\left(1 - \beta\frac{\phi}{A}\right) \qquad \text{IV.12}$$

where A is the aspect ratio a/c of an ellipsoid with principal semi-axes a and c ($a < c$). Although the derivation of IV.12 is somewhat involved, a simple interpretation of this result can be given using percolation theory.

For a heterogeneous medium, we know from chapter III that physical properties are functions of the parameter p (occupancy probability for bond or site percolation). We also know that in the case of cracks,

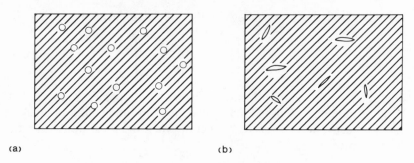

Fig. IV.11. Porous medium: (*a*) high porosity, spherical pores; (*b*) low porosity, cracks.

$p \propto Nc^3$, where N is the crack density and c the crack radius. Because the porosity is $\phi \propto Nc^2a$, we get:

$$p \propto \frac{\phi}{A}. \qquad\qquad \text{IV.13}$$

The percolation parameter p depends on two quantities: porosity ϕ and aspect ratio A. From the percolation theory point of view, it is equivalent to have a small ϕ and a small A, or large ϕ and a large A. The first situation is typical of a crack porosity and the second one of equant pores. They correspond respectively to crystalline rocks (granites for instance) and sedimentary rocks (fig. IV.11).

If we assume that K^* is a function of p, equation IV.12 can be viewed as the first term of a series development of $K^*(p)$.

$$K^* \approx K^*(0) + p\frac{\partial K^*}{\partial p} = K_1\left[1 - \beta\frac{\phi}{A}\right].$$

Here $K^*(0) = K_1$ and $\dfrac{\partial K^*}{\partial p} = -\beta K_1$. Large variations of K^* are observed near the threshold concentration p_c, that is, for large p values, whereas the above result is valid for small p values. The preceding result emphasizes the importance of pore geometry. Here the aspect ratio A is the parameter which contains microstructural information about the pores. This geometrical effect can be sufficiently strong so as to compensate for the small ϕ values.

Cracks submitted to a pressure P will close easily if they are very flat, that is, their aspect ratio A is small. In this case, the pressure dependence of the elastic modulus K^* (and other physical properties) is large. Elastic calculations show that for $P < 100$ MPa, it is possible to describe crack aperture changes by the following equation:

$$a = a_o \left[1 - \frac{2(1 - \nu^2)}{AE} P \right] \qquad \text{IV.14}$$

where a_o is the zero pressure aperture, ν Poisson's ratio, and E Young's modulus (Walsh, 1965). This equation implies that a crack of initial aspect ratio $A_o = a_o/c_o$ will close when the pressure reaches the "closure pressure"

$$P = \frac{AE}{2(1 - \nu^2)}. \qquad \text{IV.15}$$

For $A = 10^{-3}$, $E = 8 \times 10^4$ MPa, and $\nu = 0.25$, the closure pressure P_{cl} is 42 MPa.

Tensor of Elastic Stiffness Constants

The previous results are valid for an isotropic medium. Rocks often exhibit some anisotropic behavior: at large scales anisotropy results from foliations, lineations, etc., while at small scales from anisotropic minerals. Generalizing equation IV.5 (Hooke's law), one gets:

$$\sigma_{ij} = C_{ijkl} \varepsilon_{kl} \qquad \text{IV.16}$$

(there is a summation over the repeated indices k and l). The constants C_{ijkl} are the elastic stiffness constants. When deformations are adiabatic, the internal energy variation per unit volume is:

$$du = \sigma_{ij} \, d\varepsilon_{ij} \qquad \text{IV.17}$$

where $u = u(s, \varepsilon_{ij})$ is the internal energy and s the entropy per unit volume. Equations IV.16 and IV.17 express the two fundamental assumptions of linear elastic theory: stress-strain relations are linear and deformations are reversible.

A contracted matrix notation is currently more often used than the previous notation: the elastic coefficients are written C_{mn} where m and n are indices corresponding to a pair of indices ij or kl:

ij	11	22	33	23	31	12
m	1	2	3	4	5	6.

Therefore $C_{11} = C_{1111}$, $C_{44} = C_{2323}$, etc. Because of the symmetry properties

$$C_{ijkl} = C_{jikl} = C_{ijlk} = C_{jilk}$$

it is not necessary to distinguish between 31 and 13. This simplified notation system uses only six indices. Using this notation, equations 16 and 17 can be rewritten as:

$$\sigma_i = C_{ij}\varepsilon_j, \quad du = \sigma_i d\varepsilon_i.$$

However, this requires the following definition of ε_i:

$$\varepsilon_1 = \varepsilon_{11}, \quad \varepsilon_2 = \varepsilon_{22}, \quad \varepsilon_3 = \varepsilon_{33}, \quad \varepsilon_4 = 2\varepsilon_{23}, \quad \varepsilon_5 = 2\varepsilon_{31}, \quad \varepsilon_6 = 2\varepsilon_{12}.$$

There is a factor 2 on the off diagonal terms of ε_{ij} because of the symmetry ($\varepsilon_{12} = \varepsilon_{21}$, etc.). The adiabatic elastic constant tensor is:

$$C_{ij} = \left(\frac{\partial^2 u}{\partial\varepsilon_i\partial\varepsilon_j}\right)_s.$$

Because it is symmetric $C_{ij} = C_{ji}$, there are only twenty-one non-zero elastic constants. The generalized Hooke's law reduces to:

$$\begin{bmatrix} \sigma_{11} \\ \sigma_{22} \\ \sigma_{33} \\ \sigma_{23} \\ \sigma_{31} \\ \sigma_{12} \end{bmatrix} = \begin{bmatrix} C_{11} & C_{12} & C_{13} & C_{14} & C_{15} & C_{16} \\ & C_{22} & C_{23} & C_{24} & C_{25} & C_{26} \\ & & C_{33} & C_{34} & C_{35} & C_{36} \\ & & & C_{44} & C_{45} & C_{46} \\ & & & & C_{55} & C_{56} \\ & & & & & C_{66} \end{bmatrix} \begin{bmatrix} \varepsilon_{11} \\ \varepsilon_{22} \\ \varepsilon_{33} \\ 2\varepsilon_{23} \\ 2\varepsilon_{31} \\ 2\varepsilon_{12} \end{bmatrix}. \quad \text{IV.18}$$

In the isotropic case, the only non-zero components are:

$$C_{11} = C_{22} = C_{33} = K + \frac{4}{3}\mu, \quad C_{44} = C_{55} = C_{66} = \mu,$$

$$C_{23} = C_{13} = C_{12} = C_{11} - 2C_{44}.$$

A simple example of anisotropy is that of hexagonal symmetry. In this case the medium is transversely isotropic: in one particular plane all directions are equivalent. The axis perpendicular to the plane of "isotropy" is not equivalent to those lying in the plane. Hexagonal symmetry implies five non-zero constants:

$$C_{11} = C_{22} = A, \quad C_{33} = C, \quad C_{44} = C_{55} = L, \quad C_{66} = N,$$

$$C_{12} = C_{11} - 2C_{66}, \quad C_{13} = C_{23} = F.$$

FRACTURE

As long as pressure and temperature conditions are moderate, rocks deform elastically when submitted to stresses. This is only true for very

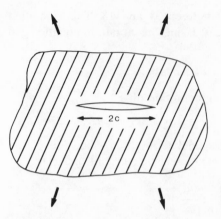

Fig. IV.12. Crack submitted to
extension stresses in a solid.

small deformations. Above a threshold stress which defines the strength
of a rock, fracture is observed. The threshold depends on the specific
situation: traction or compression. Brittle behavior results from the
existence of microcracks inside the rock. But, at sufficiently high temperatures and pressures, brittle behavior is no longer observed. For
these reasons, it is of fundamental interest to examine the stability of a
microcrack in a stressed body.

Crack Stability

Griffith (1920) analyzed the stability of an isolated crack in a solid which
was subjected to an applied stress (fig. IV.12). The stability criterion is
obtained by minimizing the total free energy of the system (cracked
rock + loading system):

$$U = -W + U_e + U_s. \qquad\qquad \text{IV.19}$$

The term W is the work done by the applied loads so that $-W$
represents the decrease in potential energy of the loading system. The
second term U_e is the strain potential energy stored in the elastic rock.
The third term is the surface energy due to the two crack surfaces which
exist within the rock. Energies W, U_e, and U_s are all expressed per unit
length of the plate.

If we assume that the applied forces are constant during deformation,
it is possible to derive the following result (Lawn and Wilshaw, 1975) for
thin plates:

$$U_e = \frac{\pi c^2 \sigma^2}{E} \quad \text{and} \quad W = 2U_e$$

where U_e is the elastic energy per unit length of plate, E the Young's
modulus, and $2c$ the crack length. Moreover:

$$U_s = 4c\gamma$$

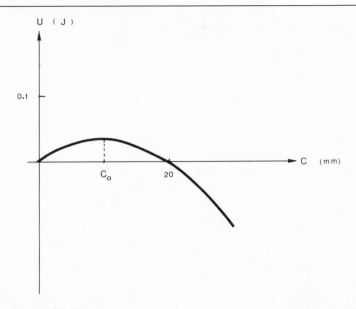

Fig. IV.13. Total energy for a crack submitted to tensile stresses (Griffith's model): $\gamma = 1.75\ Jm^{-2}$, $E = 6.2 \times 10^{10} Pa$, $\sigma = 2.6 \cdot 10^6 Pa$ (glass). The equilibrium position corresponds to $c_o = 10$ mm.

where γ is the thermodynamic surface energy. The total energy $U(c)$ is plotted in figure IV.13. At small c values, U increases almost linearly as a result of the increase in surface energy. At large c values, the mechanical energy $U_m = (U_e - W) = -\dfrac{\pi c^2 \sigma^2}{E}$ becomes the dominant term, so that U decreases with increasing c. The equilibrium length c_o is determined by the condition $\dfrac{dU}{dc} = 0$:

$$c_o = \frac{2\gamma E}{\pi \sigma^2}.$$

Cracks larger than c_o are not stable. The maximum stress which can be applied to a rock containing a crack of length $2c$ is:

$$\sigma_1 = \left(\frac{2\gamma E}{\pi c}\right)^{1/2}. \qquad \text{IV.20}$$

σ_1 depends on the intrinsic physical parameters (Young's modulus E and surface energy γ) and on an extrinsic parameter, the crack length $2c$.

Table IV.4. MECHANICAL STRENGTH OF ROCKS IN TRACTION

Rock Type	σ_1 (traction) MPa
Calcite	0.5–10
Sandstone	6–50
Granite	6–15

The mechanical strength of a rock in traction is small. Table IV.4 gives a range of observed values in the laboratory. The above values can be viewed as indirect measurements of the length of the largest crack $c_o = \dfrac{2\gamma E}{\pi\sigma^2}$ in a sample. Generally they are larger than the value reported on figure IV.13 for glass and correspond to c_o values of a few mm. This length is close to the average grain size, which suggests that cracks are localized at grain boundaries. Equation IV.20 shows that the mechanical strength in traction is not an intrinsic property: the larger the cracks are, the smaller is σ_1. The largest crack controls the mechanical response. It is thus normal to observe scale effects and to measure smaller strengths on larger samples. In fact the sample length determines the maximum possible crack length. Extrapolating, it can be concluded that mechanical strength in traction approaches zero in the limit of infinitely large samples.

Griffith's result can be generalized to any configuration by combining the three possibilities described in figure IV.14. There are only three fracture propagation modes. As before, we define the total mechanical energy of the system as $U_m = U_e - W$ and introduce the surface energy $U_s = 2\gamma c$ (crack propagation is supposed to take place in the right

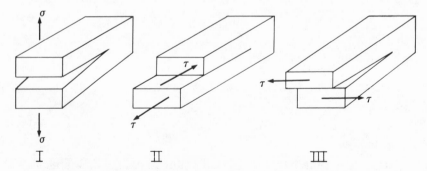

Fig. IV.14. The three modes of fracture: mode I (opening mode); mode II (sliding mode); mode III (tearing mode).

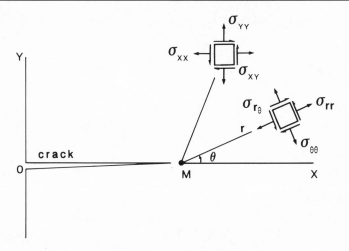

Fig. IV.15. Crack-tip stresses.

direction only). The equilibrium condition is determined through the condition:

$$\frac{d}{dc}(U_m + U_s) = 0.$$

Defining $G = -\dfrac{dU_m}{dc}$ as the crack extension force, we find:

$$G_c = 2\gamma, \qquad\qquad \text{IV.21}$$

for a linear elastic solid G is calculated from fracture mechanics either as a function of elastic and geometrical parameters on one hand, or stresses on the other hand.

The stress field at a crack tip (figs. IV.14 and 15) is $\sigma_{ij} = \dfrac{K}{\sqrt{2\pi r}}f_{ij}(\theta)$, where r and θ are the polar coordinates of a point M (Lawn and Wilshaw, 1985). In the case of homogeneous loading:

$$K_I = \sigma\sqrt{\pi c}\,(\text{mode I}), \quad K_{II} = \tau\sqrt{\pi c}\,(\text{mode II}),$$

$$K_{III} = \tau\sqrt{\pi c}\,(\text{mode III}).$$

The following results can be derived for a thin plate:

$$G_I = \frac{K_I^2}{E}, \quad G_{II} = \frac{K_{II}^2}{E}, \quad G_{III} = \frac{K_{III}^2}{E}(1 + \nu).$$

The equilibrium condition IV.21 can also be written as $K = K_c$. Using $G_I = \dfrac{K_I^2}{E}$ and $K_I = \sigma\sqrt{\pi c}$, the threshold condition $G_c = 2\gamma$ leads to $\sigma_1 = \left(\dfrac{2\gamma E}{\pi c}\right)^{1/2}$ which is Griffith's result for mode I fractures. Similar relations hold for modes II and III. When the loading is not homogeneous, equation IV.21 is always the critical equilibrium condition, but G may have a complex form.

In reality the thermodynamic surface energy γ does not adequately describe the fracture resistance. There is also a dissipative component which should be included because crack propagation is never completely reversible. To take into account this component, the fracture surface energy Γ (or fracture toughness) is used instead of γ. The values of γ are in the range 1–10 J m^{-2} (for minerals), whereas Γ values are close to 100 J m^{-2}. The differences between γ and Γ are explained by the existence of a microcracked zone at the crack tip. In the limit where the dissipative component goes to zero, Γ reduces to the ideal value γ.

There is a fundamental difference between modes II and III and mode I. This difference is that directional stability exists only in mode I. Equations IV.20 and 21 give a stability criterion, but they do not contain any information on the direction of crack extension. Generalizing Griffith's criterion, we can assume that the propagation direction will be such as to minimize the total system energy. For a propagation δc, $\delta U = -(G - 2\gamma)\delta c$ so that this corresponds to maximizing the quantity $(G - 2\gamma)$. For an isotropic situation, this reduces to maximizing the mechanical energy release rate G. The simplicity of this conclusion is misleading because when a crack extends out of its original plane, calculations are generally intractable. Only a few particular results can be obtained:

(*a*) in mode I, a plane crack propagates in its original plane ("directional stability");

(*b*) in modes II and III, a plane crack is deflected away from its original plane ("directional instability"). The propagation plane may be predicted using the above criterion: deviant cracks are deflected to a stable path of orthogonality to the greatest principal tensile stresses (fig. IV.16).

Directional instability of cracks submitted to shear loading was originally observed by Brace and Bombolakis (1963), at a time when fracture mechanics was still a very young science. The major consequence of this result is that equation IV.21 is only a criterion for initiating mode II and III cracks.

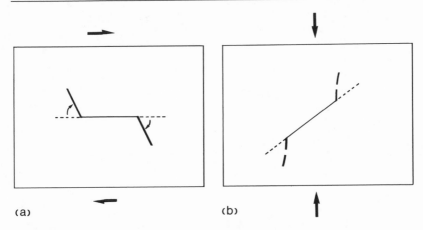

Fig. IV.16. Directional instability of mode II cracks. (*a*) pure mode II; (*b*) mixed mode.

When a rock is not isotropic, the maximum of $(G - 2\gamma)$ can be obtained by minimizing γ. If the surface energy is anisotropic, the favored orientation corresponds to that of minimum γ. A prefractured medium, which is also partially cemented, is a nice example of an anisotropic system with anisotropic γ. In this medium reactivating an old fracture can be less costly in energy than the creation of a new fracture perpendicular to the maximum tensile stresses. In this case minimizing γ is more efficient than maximizing G. This simple example explains why reactivation of old fractures (rather than creation of new fractures) has been observed during the Hot Dry Rock project at Los Alamos.

Subcritical Cracks

We have shown that the crack stability criterion (in mode I) is given by $G = G_c$ (equ. IV.21) or $K = K_c$, and have consequently assumed that cracks are arrested when $G < G_c$. In fact these assumptions ignore slow crack propagation phenomena. In the subcritical regime (i.e., $G < G_c$, $K < K_c$), very slow crack propagation is observed ($v < 10^{-3}$ m s^{-1}) in mode I for $K < K_c$.

Subcritical crack behavior can be explained by dissipative forces (friction forces) which resist crack growth. We have already considered these forces in order to explain the differences between γ (thermodynamic surface energy) and Γ (the true crack surface energy). When a crack propagates at a velocity $v = \dfrac{da}{dt}$, the energy balance is:

$$(G - 2\gamma)\delta a = F\delta t \Leftrightarrow (G - 2\gamma)\frac{da}{dt} = F$$

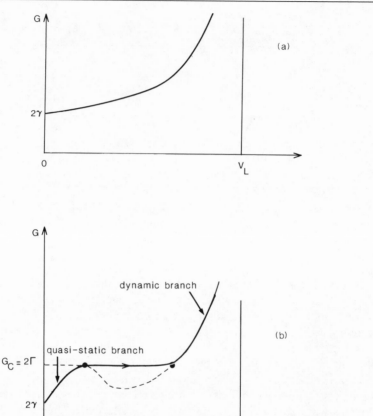

Fig. IV.17. (a) Plot of $G(v)$ in the ideal case of no friction: v_L is the limiting velocity for dynamic propagation. (b) Plot of $G(v)$ including friction. The plot can be viewed as the superposition of the ideal plot (a) and of friction effects. In the range $[0-v_c]$, subcritical crack propagation is observed, in the range $[v_R, v_L]$, dynamic propagation is observed. (From Maugis, D., 1985, Subcritical crack growth, surface energy, and fracture toughness of brittle materials, in *Fracture Mechanics of Ceramics*, Bradt, Evans, Hasselman, and Lange, ed. New York: Plenum Press).

where $F(v)$ is a function of crack velocity v and represents energy dissipation. The difference between G and 2γ is used to overcome dissipative forces.

In the absence of dissipation, $G(v)$ is an increasing function from $G = 2\gamma$ to infinity (fig. IV.17a). The existence of energy dissipation

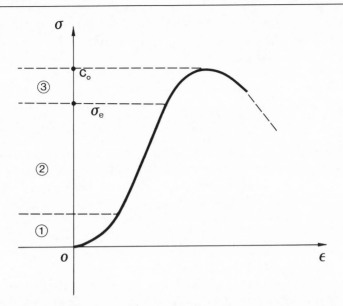

Fig. IV.18. Stress-Strain plot for a cylindrical rock sample submitted to a uniaxial compression σ. ε is the axial strain.

modifies this behavior at low v values: $F(v)$ first increases with v, but only up to a limit. Above this limit $F(v)$ decreases with increasing v. By superimposing the ideal plot of figure IV.17a and the effects of friction (which exhibits a maximum), we obtain figure IV.17b. Dissipative forces are dominant at low v values but negligible at high v values. The critical threshold $G = G_c$ is reached at $v = v_c$. Then two velocities are possible and there is a velocity jump from $v = v_c$ (critical velocity) to $v = v_R$ (dynamic rupture velocity). There are two regimes of crack propagation (or two branches on the plot): the quasistatic regime ($0 \leq v \leq v_c$) and the dynamic regime ($v_R \leq v \leq v_L$). The domain $[v_c, v_R]$ is a forbidden domain. v_c values are of the order of $10^{-3} - 10^{-2}$ m s^{-1}.

The sign of the quantity $(G - 2\gamma)$ defines that of the velocity $v = \dfrac{da}{dt}$, as noted by Rice (1978). For $G < 2\gamma$, $v < 0$: the crack closes. Crack closing is never completely reversible, because bonds are never exactly reconstructed.

Compressive Strength

Mechanical properties of a real rock are controlled by a population of cracks and not by a single crack. When a rock sample is submitted to a uniaxial or triaxial compression test, the stress-strain curve on the average is similar to that of figure IV.18. Three domains can be recognized in such a plot: (1) corresponds to the closure of cracks oriented perpendicular to the compression stress σ; (2) represents the elastic domain (the slope is the static Young's modulus in uniaxial compression); (3) corresponds to the opening and propagation of cracks oriented parallel (or subparallel) to σ. C_o is the uniaxial compressive strength. Failure begins at C_o and occurs progressively beyond this point. The post-failure curve can be recorded when stiff testing machines are used (or alternatively when servo control has a very short response time, on the order of 10 milliseconds). A detailed discussion of the post-failure curves has been presented by Paterson (1978). Note the importance of dilatancy during phase (3): cracks are opened and the total volume does not decrease as classical elasticity would predict $\left(\dfrac{\Delta V}{V} = \dfrac{(1 - 2\nu)}{E} \sigma \text{ for elastic uniaxial compression} \right)$. During dilatancy, the physical properties of the rock are modified (variation of V_p and V_s, variation of electrical conductivity, etc.).

Fracture under compression becomes a problem when there is a high crack density. Crack initiation criteria, such as $G = G_c$, are not fracture criteria in this case. In mode II and III, cracks are deflected away from their plane. Thus their propagation direction depends on the local stress field. This stress field itself depends on the crack geometry and on its neighbors. Approximate solutions to this complex problem of interacting cracks have been suggested by various authors (see Guéguen et al., 1990). Kemeny and Cook's model (1987) is presented in figure IV.19: each crack propagates by nucleating two wing cracks. The angle β is obtained by maximizing K_I. Crack interactions are considered either between axially aligned cracks (axial fracture, IV.19c) or between columns of cracks (shear fracture, IV.19d).

Percolation theory can also be used to examine this problem. Koslenikov and Chelidze (1985) were the first to apply it to fracture by considering that fracture takes place when the quantity Nc^3 (where N is the crack number density and c the crack radius) reaches a critical value. Fragmentation and fracture are then two, very distinct, possibilities. Fragmentation occurs when crack nucleation is homogeneous (fig. IV.20d) and crack interactions are negligible. Crack interactions and anisotropy produce local crack concentrations which coalesce to form a macroscopic fracture (fig. IV.20e). Since cracks are anisotropic objects which interact through their stress fields, it is in general the second possibility which is observed. In some particular circumstances, fragmentation can take place: for instance when a stress pulse produces homogeneous crack nucleation (explosion) or when local stress concentrations exist everywhere within the rock (highly porous rocks). Statisti-

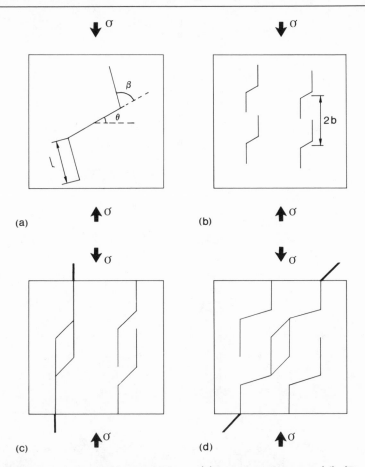

Fig. IV.19. Development of axial fractures (c) and shear fractures (d). (From Kemeny, J. M., and N.G.W. Cook, 1987, Crack models for the failure of rocks in compression, in Proc. 2nd Int. Conf. on Constitutive laws for Engineering Materials, Tucson, Arizona.

cal models of fracture using percolation and fractal concepts have been presented by Hermann and Roux (1990) for disordered media.

Mechanical strength of rocks under compression is higher than under traction, as can be seen by comparing tables IV.4 and IV.5. If σ and τ are the normal and shear stress on a fracture plane, Mohr's envelope is the curve which describes the maximum possible values of σ and τ. Assuming that the cracks in a rock are randomly oriented, it is then possible to show that Mohr's envelope is given by the plot shown in figure IV.21. At small σ values (negative values correspond to tensile

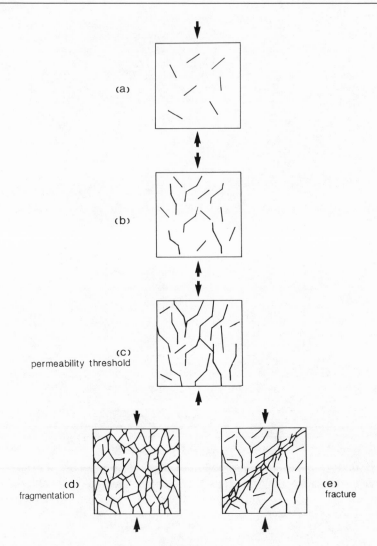

Fig. IV.20. Development of cracks by homogeneous and isotropic nucleation: case *d*; or by non-homogeneous and anisotropic nucleation (interactions): case *e*. The permeability threshold is reached before fragmentation or fracturing.

stresses[1]) the envelope is a parabola: $\tau^2 - 4T_o\sigma = 4(T_o)^2$ where T_o is the mechanical strength in traction (Griffith's criterion). At high σ values, cracks are closed and internal friction on crack surfaces must be

---

[1] The sign convention used here is that of rock mechanics. It is the opposite of that of elasticity. Tensile stress are positive in elasticity and negative in rock mechanics.

Table IV.5. MECHANICAL STRENGTH OF ROCKS IN COMPRESSION

Rock Type	C_o (compression) MPa
Calcite	7–60
Sandstone	7–50
Granite	60–180

taken into consideration. Then the envelope is a straight line: $\tau = S_o + \mu\sigma$, where μ is the friction coefficient (Coulomb criterion).

A complete analysis of the failure criteria is given in the text by Jaeger and Cook (1978).

Friction

Rocks usually contain old fractures that are more or less cemented. When cementation is sufficient, the rock is similar to an intact rock for which surface energy γ is anisotropic: γ is smaller in the plane of a former fracture. Healing is never perfect however, even in the case of glass where γ is approximately five times lower than its initial value. When cementation is limited or negligible, the above description does not hold. Mechanical strength is then controlled by the sliding of one fracture surface on the other, which itself is controlled by the rugosity of the fracture surface (fig. IV.22). Experimental friction laws have been

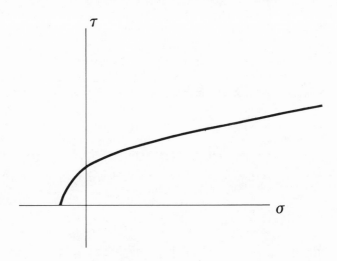

Fig. IV.21. Mohr's envelope in the plane (τ, σ): τ is the shear stress and σ the normal stress on the fracture plane.

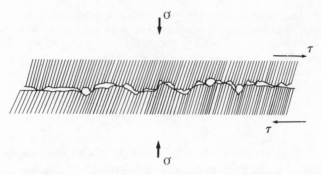

Fig. IV.22. Roughness of real fracture surfaces.

obtained by Byerlee:

$$10 \text{ MPa} \leq \sigma \leq 200 \text{ MPa}, \quad \tau = 0.85\sigma$$
$$200 \text{ MPa} \leq \sigma \leq 1500 \text{ MPa}, \quad \tau = 50 + 0.6\sigma. \qquad \text{IV.22}$$

τ and σ are respectively the shear stress and normal stress on the fracture plane. It is interesting and surprising that Byerlee's laws IV.22 hold for any rock type. There is no satisfactory theory to explain these friction laws.

Friction can be attributed to rough surfaces. During slip, asperities break and there is a mutual indentation of each fracture surface by the other. This leads to the production of a gouge zone composed of crushed debris. If the gouge zone is large, its properties can control the mechanical response of the system. The ultimate result depends strongly on the hardness of the minerals that are present. When the hardness of the minerals involved in this process is high (quartz, feldspars), stick slip takes place instead of stable slip. Similar behavior is observed when confining pressure is increased (fig. IV.23). Temperature on the other hand has a reverse affect: increasing T favors stable slip. Ductile minerals (such as shales) also have a lubricating effect that stabilizes slip. Oscillations observed in the stick slip regime are an example of relaxation oscillations. The system has two equilibrium positions corresponding to two friction coefficients: a static friction coefficient μ and a dynamic one $\mu - \Delta\mu$ (fig. IV.24). This model is frequently used for earthquakes and slip along faults.

PLASTICITY

When confining pressure and temperature are sufficiently high, rocks submitted to stresses deform plastically and can develop large strains. In such a case a finite strain develops and not an infinitesimal one. We

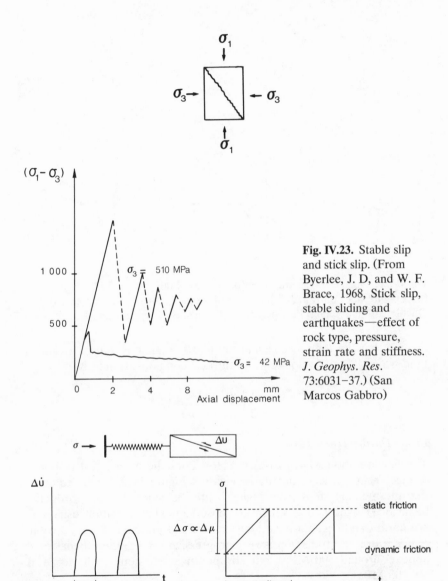

Fig. IV.23. Stable slip and stick slip. (From Byerlee, J. D, and W. F. Brace, 1968, Stick slip, stable sliding and earthquakes—effect of rock type, pressure, strain rate and stiffness. *J. Geophys. Res.* 73:6031–37.) (San Marcos Gabbro)

Fig. IV.24. Model of relaxation oscillation. Slip occurs during a time $t_l(\Delta u \neq 0)$. The spring is then loaded again up to the static friction stress.

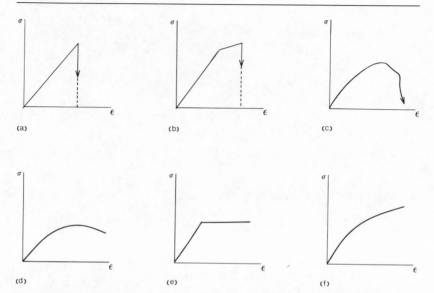

Fig. IV.25. Ductile-fragile transition on a stress-strain plot. (*a*) Highly fragile; (*b*) Fragile; (*c*) Controlled failure; (*d*) Moderately ductile; (*e*) Ductile; (*f*) Ductile with hardening. Parameters (*P, T*) increase from (*a*) to (*f*).

assume that the rock behavior in the plastic regime is that of a viscous fluid which flows at velocity $\vec{v}$. Strain rate components are given by

$$\dot{\varepsilon}_{ij} = \frac{1}{2}\left(\frac{\partial v_i}{\partial x_j} + \frac{\partial v_j}{\partial x_i}\right).$$

Brittle-Ductile Transition

The first question which arises is "How does the mechanical response change from brittle to ductile behavior?" Figure IV.25 shows some of the intermediate steps from highly brittle behavior (IV.25a) to ductile behavior (IV.25e and f) which are observed. This transition usually is not abrupt and occurs over a temperature and pressure range. At room temperature, the transition is very progressive for a granite (behavior *d* appears around 300 MPa), but abrupt for a limestone (behavior *e* at 50–100 MPa).

These results can be understood if one realizes that two microscopic mechanisms can take place at the same time: crack propagation and dislocation propagation. Dislocations are line defects within crystals, at the atomic scale (see below). They are stable (frozen in) when temperature is low and mobile at high temperatures. As we have seen, cracks are stable as long as the stress does not exceed a critical value σ_1.

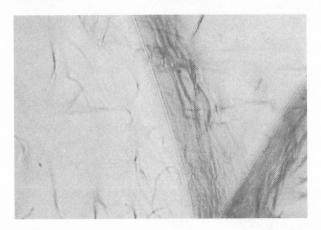

Fig. IV.26. Crack-dislocation interactions in an olivine crystal at 900°C. The crack is trailing the dislocations. (From Darot, M., Y. Guéguen, and Z. Bencheman, 1985, Ductile-brittle transition investigated by micro-indentation: results from quartz and olivine. *Phys. Earth and Plan. Int.* 40 (1985):180–86.)

Beyond this threshold, crack propagation occurs and fractures develop. The critical value σ_1 depends on one geometrical parameter (crack length) and two physical parameters (E and γ) as can be seen in equation IV.20. When pressure and temperature conditions are changed, E and γ vary. Changes in E are small compared to that of γ: between 0°C and 1000°C, E decreases for most minerals by less than 50%. In fact Γ, which includes dissipation, should be used instead of γ as discussed previously. But it is precisely the dissipative processes which increase with temperature. For these reasons, σ_1 increases with T. The reverse is true for σ_1', the threshold for dislocation mobility. Consequently increasing T favors dislocation propagation and makes crack propagation less likely. There is a progressive transition from a state where cracks are unstable and dislocations frozen in (fig. IV.25a) to a situation where dislocations are mobile and cracks stable (fig. IV.25e–f).

One of the reasons why Γ increases is the dislocation mobility itself. Dislocations are always present near cracks. When the temperature is sufficiently high, these dislocations are put into motion by the stress field of the mobile crack. The forces due to the crack do plastic work in displacing the dislocation segments as the crack advances (fig. IV.26). By this process the fracture resistance term Γ increases. Both processes, crack propagation and dislocation propagation, are interdependent. There exists a domain, called the transition domain, when both are simultaneously active.

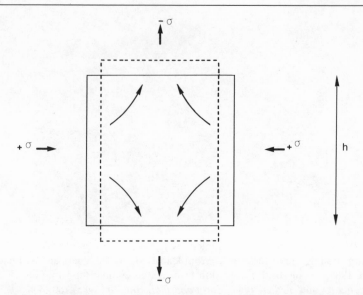

Fig. IV.27. Diffusion creep: atoms are moving from faces submitted to compression toward faces submitted to extension. The result of atomic diffusion is a finite strain. Compression stresses are positive (rock mechanics convention).

Pressure effects can be readily understood in terms of the microscopic processes. High confining pressures tend to close cracks and suppress crack propagation (a minimum crack opening is required). But confining pressure has a negligible effect on dislocations.

Diffusion and Dislocations

In the ductile regime, the deformation of dry rocks results from three distinct processes: diffusion creep, dislocation creep, and superplastic creep. We will use the term creep due to its common usage. A creep test is an experiment where stress σ is held constant and the strain ε is recorded as a function of time. The three processes we have mentioned correspond to three different scales of microscopic discontinuities (Nicolas and Poirier, 1976).

Diffusion creep may be important for very fine grained rocks at high temperatures (near the melting temperature T_m). Under those conditions, atoms are mobile enough to allow a non-negligible deformation through bulk diffusion from zones in compression toward zones in extension (fig. IV.27). Fick's law states that the particle flux J (atoms or vacancies) is proportional to the concentration gradient:

$$J = -D_v \frac{\partial c}{\partial x}. \qquad\qquad \text{IV.23}$$

D_v is the bulk diffusion coefficient of vacancies and c the number of vacancies per unit volume. Let us assume that atom migration in one direction is the result of vacancy migration in the opposite direction and that the vacancy concentrations are different on the two opposite faces (fig. IV.27):

$$C_1 = C_o \exp(-\sigma v/RT) \quad \text{(face in compression)}$$
$$C_2 = C_o \exp(\sigma v/RT) \quad \text{(face in extension)}.$$

(C_o = equilibrium concentration when $\sigma = 0$; v = atomic volume; $R = 8.3$ J mole^{-1}°K^{-1}). The average vacancy concentration gradient is $\dfrac{C_2\,C_1}{h}$. Then the vacancy flux between two adjacent faces is (using equ. IV.23)

$$J_v = -D_v\left(\frac{C_2 - C_1}{h}\right).$$

The number of vacancies which are transported from one face to the other per second is

$$N = -J_v h^2 = D_v(C_2 - C_1)h.$$

Transport of a complete atomic layer of thickness b results in a strain

$$\Delta e = \frac{b}{h}.$$

Such a layer contains $\dfrac{bh^2}{b^3}$ vacancies. Consequently, when N vacancies are transported, the strain is $\dfrac{\Delta e}{bh^2/b^3}N = \dfrac{b^3}{h^3}N$. When this strain is the strain produced in one second, it is called the strain-rate $\dot{\varepsilon}$. Using $v = b^3$, $\bar{c} = C_o v$ (volumetric concentration), we find: $\dot{\varepsilon} = \dfrac{b^3}{h^3}N = \dfrac{Nv}{h^3}$. Assuming that the stress σ is low and temperature T is high, we can approximate N:

$$N = D_v(C_2 - C_1)h \approx \frac{2\sigma D}{RT}h$$

because $\sigma v \ll RT$ and by definition $D = \bar{c}D_v$. Using this simplified model, the strain rate for diffusion creep becomes

$$\dot{\varepsilon} = \frac{2\sigma v}{RT}\frac{D}{h^2} \qquad\qquad \text{IV.24}$$

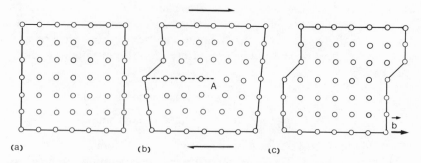

Fig. IV.28. Glide on an atomic plane in a cubic crystal. Initial state = (a); Intermediate state = (b); Final state = (c). State (b) shows an edge dislocation at point A. The edge dislocation is perpendicular to the plane.

$\dot{\varepsilon}$ is a linear function of σ and varies as h^{-2}. Moreover $\dot{\varepsilon}$ depends strongly on T because D is an exponential function of T: $D = D_o\exp(-E/RT)$. This explains why diffusion creep becomes important at very high temperatures (where D is high) and for very small grain sizes (h^{-2}). Both of these conditions are in general not met in the crust (primarily due to low T) and diffusion creep probably does not play any role in this part of the earth.

The same conclusion does not hold for *dislocation creep*. We have previously discussed the important role of dislocations for the brittle-ductile transition. Dislocations are atomic line defects and become mobile at moderate temperature ($T < 0.5T_m$). For the simple case of a monatomic cubic crystal (fig. IV.28), the slip process propagates progressively in the crystal: between the initial state (undeformed crystal) and the final state (deformed crystal), there are intermediate states. At a given time, slip extends up to a limit, the dislocation line (figs. IV.28 and 29). This line is a discontinuity through which the displacement vector $\vec{u}$ experiences a jump of $\vec{b}$. The vector $\vec{b}$, called the Burgers' vector, describes the amount and direction of slip. The above process can be understood if one notes that progressive displacements of the dislocation line break only one atomic bond at each step (fig. IV.30). An instantaneous displacement of the upper half of the crystal by $\vec{b}$, while the lower half remains fixed, would simultaneously break many bonds and for that reason is energetically improbable. Figures IV.28 and IV.30 represent "edge" dislocations: in this case $\vec{b}$ is perpendicular to the dislocation line. An edge dislocation appears as the limit of a wedge (or an additional atomic plane). Figure IV.29 shows that "edge" dislocations can be transformed into "screw" dislocations. In the first case $\vec{b}$ is perpendicular to the line, and in the second case it is parallel to it.

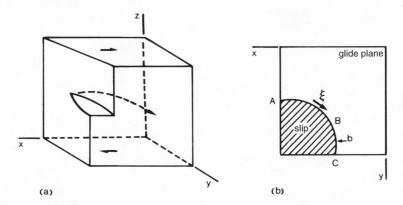

(a)　　　　　　(b)

Fig. IV.29. A curved dislocation line which divides the glide plane in two areas: one has already experienced slip of $\vec{b}$. $\vec{\xi}$ is the unit vector tangent to the dislocation and varies along the line. $\vec{b}$ is the Burgers' vector and is constant. (*a*) Global view; (*b*) Section in the glide plane.

As for diffusion creep, we can calculate the strain rate $\dot{\varepsilon}$ for the case of dislocation creep. Consider a dislocation (fig. IV.31) which is at point A. When this dislocation has crossed the crystal, a shear strain $\varepsilon = \dfrac{b}{h}$ is produced. We assume that for a small displacement $\delta x = AA'$, a strain $\delta\varepsilon = \dfrac{b}{h}\dfrac{\delta x}{l}$ is produced. If δt is the time needed for the displacement δx, the glide velocity V is $V = \delta x / \delta t$, and the strain rate is

$$\dot{\varepsilon} = \frac{b}{hl}V.$$

Extending this result to a distribution of N dislocations over an area $S = hl$, and introducing the dislocation density $\rho = \dfrac{N}{S}$, one finds:

$$\dot{\varepsilon} = \rho b V \qquad\qquad \text{IV.25}$$

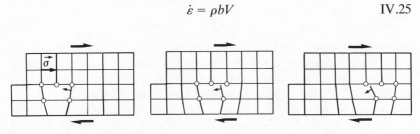

Fig. IV.30. Progressive slip by dislocation displacement. Each unit step corresponds to switching a bond.

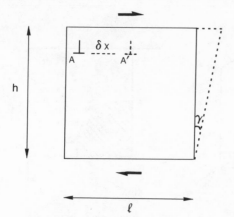

Fig. IV.31. Shear resulting from dislocation glide. γ is the shear angle when $\delta x = 1$.

which is the Orowan equation. It is a transport equation which states that strain rate is the product of three quantities: (density of strain carriers) × (charge of carriers) × (velocity of carriers). In the present case, a dislocation "carries" a displacement b. The density ρ depends on stress, and the velocity V is a function of temperature T and pressure P. Equation IV.25 leads in general to creep laws of the following type:

$$\dot{\varepsilon} = A\,\sigma^{\,n}\exp\left(-\frac{E}{RT}\right) \qquad\qquad \text{IV.26}$$

where E is the activation energy per mole or per molecule. In the first case (mole) equation IV.26 is the appropriate one. In the second case (molecule), E_a/k is used instead of E/R where $E = NE_a$ and $R = Nk$ ($k = 1.38\ 10^{-23}$ J $°K^{-1}$ is Boltzmann's constant and $N = 6.02\ 10^{23}$ is Avogadro's number).

Dislocation creep laws are in general non-linear ($n \approx 2\text{--}4$). Dislocation glide is often controlled by diffusion, which explains why E is close to the bulk diffusion activation energy. Indeed when dislocations are progressing through glide (fig. IV.30), they stay in their initial plane as shown in figure IV.32. This plane is defined by two directions: the Burgers' vector and dislocation line. However, if there are obstacles to their displacement in this plane, they may need to leave the original plane in order to bypass them. Such a displacement away from the glide plane is called dislocation climb. Climb is non-conservative, that is, a volume element is created by the dislocation displacement. An edge dislocation climbs either by absorbing vacancies or interstitial atoms on the extra half plane. This requires diffusion and explains why dislocation creep is often diffusion controlled (fig. IV.33). Table IV.6 gives

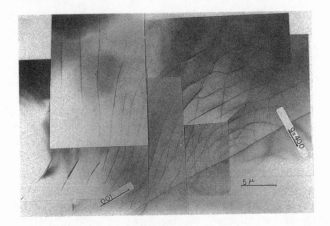

Fig. IV.32. Dislocations in their glide plane near a subgrain boundary. Transmission Electron micrograph of an olivine crystal. Glide plane (010); Burgers' vector [100]; Subgrain plane (100).

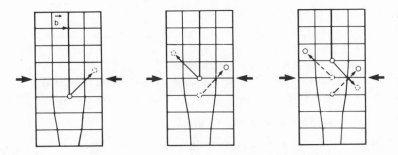

Fig. IV.33. Diffusion controlled climb of edge dislocations: the process is controlled by atomic diffusion.

Table IV.6. MEASURED VALUES OF DISLOCATION CREEP PARAMETERS

	A	n	$E_a(eV)$
Quartz SiO$_2$	5×10^6	3	2.0
Forsterite Mg$_2$SiO$_4$	10^6	2.6	4.7

Source: W. F. Brace and D. C. Kohlstedt, 1980. Limits on lithospheric stress imposed by laboratory experiments. *J. Geophys. Res.* 89:6248–52, and Guéguen, Y. and M. Darot, 1982, Upper mantle plasticity from laboratory experiments. *Physics of the Earth and Planetary Interiors* 29:51–57.
Note: $1eV = 1.6 \times 10^{19}$ Joules.

some experimentally measured values of parameters A, n, E_a for dislocation creep in two important minerals: quartz and olivine (forsterite). Stresses are in MPa.

Superplastic creep is a special case. Some laboratory experiments as well as observations on naturally deformed rocks suggest that this deformation regime could be locally important in the crust. Superplastic creep occurs when grain size is very small and grain boundary sliding is important. Grain boundary sliding is diffusion accommodated, so that the creep law is similar to that given by equation IV.24. The pre-exponential factor is however larger by a factor of at least 50.

Extrapolation of laboratory data to the crust is difficult. Geological strain rates $(10^{-13}-10^{-14}s^{-1})$ are very small as compared to laboratory strain-rates $(10^{-3}-10^{-8}s^{-1})$. It seems however that for dry rocks, dislocation creep and superplastic creep may be of some importance only in the lower crust, or locally in highly stressed zones submitted to high temperature and high pressure conditions (shear zones, mylonites, etc.).

PROBLEMS

IV.a. Change of Coordinate Systems

Let $\vec{\sigma}(\vec{n}) = \sigma_{ij} n_j \vec{e}_i = \sigma'_{ij} n'_j \vec{e}''_i$ be the stress vector on a plane defined by its normal vector $\vec{n}$. Unit vectors $\vec{e}_i$ and $\vec{e}''_i$ correspond to two different Cartesian coordinate systems such that $\vec{e}''_i = \alpha_{ij} \vec{e}_j$, where $P = (\alpha_{ij})$ is the matrix which describes the change of coordinates. Calculate the quantity $\vec{\sigma} \cdot \vec{n}$. Using $\sigma'_{ij} = \vec{e}''_i \vec{\sigma}(\vec{e}''_j)$ show that $\sigma'_{ij} = \alpha_{ik} \alpha_{jl} \sigma_{ke}$.

IV.b. Principal Stresses and Mohr's Circle

In the Cartesian coordinate system $Oxyz$, the stress tensor σ_{ij}

$$= \begin{bmatrix} \sigma_{11} & \sigma_{12} & 0 \\ \sigma_{12} & \sigma_{22} & 0 \\ 0 & 0 & 0 \end{bmatrix}.$$

(1) Calculate the principal stresses σ' and σ''. Calculate the angle θ between Ox axis and the direction of σ'.

(2) Assuming that $\sigma_{12} = 0$, show that the normal stress σ and the shear stress τ on a plane defined by a normal $\vec{n}$ in the Oxy plane are:

$$\sigma = \frac{\sigma_{11} + \sigma_{22}}{2} + \frac{\sigma_{11} - \sigma_{22}}{2} \cos 2\varphi, \quad \tau = -\frac{\sigma_{11} - \sigma_{22}}{2} \sin 2\varphi$$

where φ is the angle between the x-axis and $\vec{n}$. Show that in the (τ, σ) plane, the above equations define a circle (Mohr's circle).

IV.c. Compatibility Equations

By deriving the strain components ε_{ij} in two ways, show that the ε_{ij} must obey the following equations:

$$\frac{\partial^2 \varepsilon_{xx}}{\partial y\, \partial z} = \frac{\partial}{\partial x}\left[-\frac{\partial \varepsilon_{yz}}{\partial x} + \frac{\partial \varepsilon_{zx}}{\partial y} + \frac{\partial \varepsilon_{xy}}{\partial z} \right] \quad \text{and} \quad \frac{\partial^2 \varepsilon_{xx}}{\partial y^2} + \frac{\partial^2 \varepsilon_{yy}}{\partial x^2} = 2\frac{\partial^2 \varepsilon_{xy}}{\partial x\, \partial y}$$

(using indices permutations, there are in fact 2×3 equations).

IV.d. Lamé Constants

Hooke's law can be written using constants λ and μ instead of K and μ:

$$\sigma_{11} = (\lambda + 2\mu)\varepsilon_{11} + \lambda \varepsilon_{22} + \lambda \varepsilon_{33}$$

(and similar relations for σ_{22} and σ_{33} using permutation of the indices). Express λ as a function of K and μ.

IV.e. Interatomic Potential

Near the equilibrium position, the interatomic potential (fig. IV.10) can be approximated by:

$$U = -U_o + \frac{1}{2}kx^2 - lx^3$$

where $x = r - r_o$ is a small displacement from equilibrium and $l > 0$.

(1) Assuming $l = 0$, show that this problem is the classical harmonic oscillator problem and calculate the oscillator frequency.

(2) Assuming $l \neq 0$, and considering that at temperature T, $U = -U_o + \frac{3}{2}RT$ (for one mole, with $R = 8.3$ J mole^{-1}°K^{-1}), show that the anharmonic term lx^3 describes a dilation. What is the new equilibrium value $r_o'(T)$? Compute an expression for the thermal expansion coefficient $\alpha = \frac{3}{r_o'}\frac{\partial r_o'}{\partial T}$.

IV.f. Screw Dislocation

Consider a cylinder of radius r containing a screw dislocation lying along its axis Oz. Let b be the Burgers' vector of the dislocation. When unrolled, the cylinder containing the screw dislocation appears as it is sheared from a rectangle into a parallelogram (without the dislocation, it would be a rectangle).

(1) Show that $\varepsilon_{\theta z} = \dfrac{b}{4\pi r}$ (r and θ are polar coordinates in a plane perpendicular to Oz).

(2) What are the components of the tensors ε_{ij} and σ_{ij}?

REFERENCES

Atkinson, B. K. 1989. *Fracture Mechanics of Rock*. New York: Academic Press.

Brace, W. F., and E. G. Bombolakis. 1963. A note on brittle crack growth in compression. *J. Geophys. Res.* 68:3709–13.

Fung, Y. C. 1965. *Foundations of Solid Mechanics*. New York: Prentice Hall.

Guéguen, Y., Reuschlé, T., and M. Darot. 1990. In *Deformation Processes in Minerals, Ceramics and Rocks*, Barber, D. and P. Meredith ed. London: Unwyn, Hyman.

Henyey, F. S., and N. Pomphrey. 1982. Self-consistent elastic moduli of a cracked solid. *Geophys. Res. Letters* 9:903–6.

Hermann, H. J., and S. Roux. 1990. *Statistical Models for the Fracture of Disordered Media*. Amsterdam: North Holland.

Jaeger, J. C., and N. G. Cook. 1979. *Fundamentals of Rock Mechanics*. London: Chapman and Hall.

Koslenikov, Y., and T. C. Chelidze. 1985. The anisotropic correlation in percolation theory. *J. Phys. A.* 18:L 273.

Landau, L., and E. Lifchitz. 1967. *Theory of Elasticity*. Moscow: Editions Mir.

Lawn, B. R., and T. R. Wilshaw. 1975. *Fracture of Brittle Solids*. Cambridge University Press.

Nicolas, A., and J. P. Poirier. 1976. *Crystalline Plasticity and Solid State Flow in Metamorphic Rocks*. New York: Wiley.

Paterson, M. S. 1978. *Experimental Rock Deformation. The Brittle Field*. Berlin: Springer-Verlag.

Rice, J. R. 1978. Thermodynamics of the quasi-static growth of Griffith cracks. *J. Mech. Phys. Solids* 26:61–78.

Walsh, J. B. 1965. The effect of cracks on the compressibility of rocks. *J. Geophys. Res.* 70:381–89.

V. Circulation of Fluids: Permeability

ON THE LARGE SCALE, fluid flow is the process which assures long-range transport of material through the porous crust. Hydrothermal systems, mineral deposits, geothermal energy, the underground storage of fluids are all fields of interest for which the permeability plays a key role. The circulation of fluids also plays an important role in the mechanical and seismic response of the crust.

Rock permeability is a measure of how easily fluid flows through a rock. It is probably the most important physical parameter with which we are concerned in the Earth's crust. A petroleum reservoir with a low permeability is a very poor reservoir!

DARCY'S LAW AND PERMEABILITY

Permeability, from the point of view of physics, is a transport property. The coefficient of permeability relates a flux (the fluid flux) to a force (the fluid pressure gradient). Under normal conditions, the fluid flux is proportional to the pressure gradient, and this linear law greatly simplifies the problem.

Darcy's Law

Consider a porous, permeable media traversed by a fluid in the $+x$ direction. The "Darcy velocity" q of a fluid is defined as the fluid volume that crosses a section S perpendicular to the x-axis per unit area and per unit time: the volume is $q \cdot S$ (fig. V.1). If the viscosity of the fluid is η and the pressure gradient $\dfrac{dP}{dx}$, Darcy's law states

$$q = -\frac{k}{\eta} \frac{dP}{dx}.$$ V.1

Darcy's velocity is a volume flux. It is not the real velocity of the fluid. Equation V.1 is a linear law of transport analogous to Ohm's Law. It is not difficult to show that k has the dimensions of (length)^2:

$$[k] = [q][\eta]\left[\frac{dx}{dp}\right] = [m \cdot s^{-1}][Pa \cdot s][m \cdot Pa^{-1}] = [m^2].$$

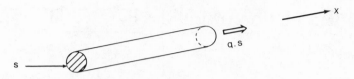

Fig. V.1. Volume flux $(q \cdot s)$ of fluid in the x-direction. q is the Darcy velocity and s the cross-sectional area.

k is the permeability coefficient: it is an effective cross-section for flow. In the general case of anisotropic media, equation V.1 becomes:

$$q_\alpha = -\frac{k_{\alpha\beta}}{\eta}\frac{dp}{dx_\beta}.$$

Darcy's velocity q can be related to the real, average velocity v of the fluid within the pores. This relation is the Dupuit-Forcheimer law:

$$q = v \cdot \phi. \qquad\qquad V.2$$

To see how this occurs note the total fluid volume Q that crosses a section S_o per second is the sum of the volume fluxes flowing through each channel (fig. V.2):

$$Q = q \cdot S_o = \sum_{i=1}^{N}(v_i s_i) = v\sum_{i=1}^{N}s_i.$$

Here we have assumed that each channel has a cross-section s_i and that $v_i = v$, for every channel i. If the porosity is distributed isotropically, $\sum_{i=1}^{N}s_i = \phi S_o$, resulting in equation V.2.

Often the unit *Darcy* is used for the permeability: $k = 1$ Darcy (1 D) when $q = 1$ cm s^{-1}, for a gradient of $\frac{dP}{dx} = 1$ atmosphere cm^{-1} and a fluid of viscosity $\eta = 10^{-2}$ poise: thus 1 Darcy $= 0.97\ 10^{-12}$ m^2. A good aquifer is one whose permeability is equal to or greater than 1 D; a

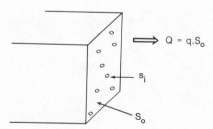

Fig. V.2. The flow of fluid through N capillaries of individual cross-section s_i. The real average velocity to the fluid in the capillaries is $v = q/\phi$.

good petroleum reservoir is one whose permeability is equal to or greater than 100 mD (10^{-1} D).

Darcy's law accurately describes the longtime fluid movement, when the real velocities are not too large. When the real velocities exceed a critical value, the approximation represented in V.1 is no longer accurate. This limit can be determined by examining the fundamental equations governing the mechanics of fluids. For a viscous fluid, the stress-velocity relationships are described by (see Landau and Lifshitz, *Fluid Mechanics*)

$$\sigma_{ij} = -p\,\delta_{ij} + \eta\left(\frac{\partial v_i}{\partial x_k} + \frac{\partial v_k}{\partial x_i}\right). \qquad \text{V.3}$$

Equation V.3 is the constitutive equation for a viscous fluid and is analogous to equations IV.6 which were used for an elastic solid. If we assume that the fluid is incompressible, that is, the density is constant, then the velocity $\vec{v} = \dfrac{\partial \vec{u}}{\partial t}$ (where $\vec{u}$ is the displacement) satisfies the equation

$$\frac{\partial v_k}{\partial x_k} = 0 \qquad \text{V.4}$$

because no change in volume, $\dfrac{\partial u_k}{\partial x_k} = 0$, implies $\dfrac{\partial v_k}{\partial x_k} = 0$. Equations V.3 and V.4 can be combined with the fundamental equilibrium equation IV.2 (assuming no body forces but a resulting non-zero acceleration) to arrive at the Navier-Stokes equation.

$$-\frac{\partial p}{\partial x_i} + \eta\,\nabla^2 v_i = \rho\frac{dv_i}{dt}. \qquad \text{V.5}$$

The inertial term $\rho\dfrac{dv_i}{dt}$ in V.5 is negligible, when

$$\left| \rho\frac{dv}{dt} \right| \ll |\eta\,\nabla^2 v|.$$

If l is the characteristic length, then l/v is the characteristic time and the two quantities are of magnitude $\left(\rho\dfrac{v}{l/v} \right)$ and $\left(\eta\dfrac{v}{l^2} \right)$, respectively. Thus the condition

$$\frac{\rho v l}{\eta} \ll 1 \qquad \text{V.6}$$

implies that the term $\rho\dfrac{dv}{dt}$ is negligible. The quantity $\dfrac{\rho v l}{\eta}$ is the Reynolds number R_e. The values ρ and η are known properties of the fluid, and the characteristic length l is fixed by the pore dimensions which serve as channels of fluid movement. When $R_e \ll 1$, the inertial term in V.4 is negligible and the Navier-Stokes equation simplifies to

$$\eta \nabla^2 v_i = \frac{\partial p}{\partial x_i}.$$

Formally, one can view this equation as a relation between velocity (or flux) and the pressure gradient (force). Darcy's law corresponds to a linear solution which approaches the solution of Navier Stokes when $R_e \ll 1$. This condition is usually satisfied in porous media. For very high permeabilities and/or very low viscosities the velocity will be very high, and $R_e \gg 1$: fluid motion is no longer laminar but turbulent and Darcy's law is not applicable. This often occurs for the flow of gases because their viscosities are extremely low. For example, the Reynolds number for Nitrogen N_2 in 10 μm pores is

$$\rho \approx 1 \text{ kg m}^{-3}, \qquad \eta \approx 10^{-5}, \quad l \approx 10^{-5} \Rightarrow R_e \approx v.$$

Thus Darcy's law would not be applicable for velocities greater than 1 m s^{-1}. In this regime of high inertial forces, Darcy's law is replaced by the law of Forcheimer

$$\frac{dp}{dx} = a\eta q + b\rho q^2.$$

Another case where Darcy's law is not applicable occurs when the pressure (density) of a gas is very low. In this case the mean free path of the molecules becomes greater than the pore dimensions. The gas cannot be considered a continuous medium and the mechanics of fluids is not applicable: this is the Klinkenberg effect. In this case the apparent permeability depends on the average pressure and can be described by Klinkenberg's law:

$$k_{\text{app}} = k\left(1 + \frac{\alpha}{p}\right).$$

Permeability of Rocks

A permeability of 1 darcy (1 D) is a high permeability. Permeabilities greater than 1 D are only encountered in gravels (10^3 D or greater) and sandy gravels (10 D or greater). In rocks of more general concern to us, the permeability varies greatly and depends on the nature of the rock. Plutonic rocks in general have very low porosities and permeabilities

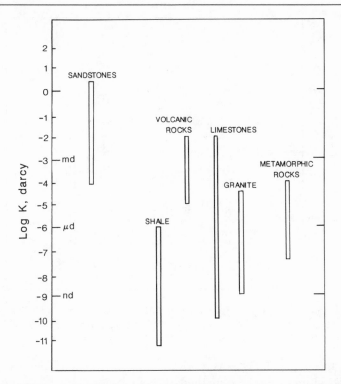

Fig. V.3. Laboratory-measured permeabilities. Hydrostatic pressure < 10 MPa, $T = 25°$ C. (After Brace, W. F., 1980, Permeability of Crystalline and Argillaceous rocks, *Int. J. Rock Mech. Min. Sci. and Geomech. Abstr.* 17:241–51.)

($k < 10$ μD). Volcanic rocks on the other hand, usually have much higher porosities and permeabilities ($k > 1$ mD). Sedimentary rocks cover a wide range of permeabilities: from very low values (argillaceous sediments, $k < 1$ μD) to much higher values (sands, $k \sim 1$ D). Sandstones and carbonates are of particular interest because they form the principal petroleum reservoirs. Figures V.3 and V.4 show laboratory measurements on various materials compiled by Brace (1980 and 1984). Very low permeabilities can occur in granites, carbonates, and shales. Carbonates in particular show an extremely wide variability in values (9 orders of magnitude).

Laboratory measurements, on samples with sufficiently large permeabilities, are based on the measured volume flow:

$$k = \frac{\text{volume flux} \times \eta}{S \times \dfrac{dp}{dx}}.$$

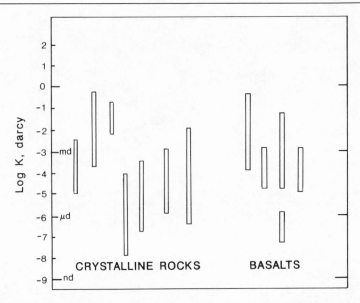

Fig. V.4. Permeabilities measured in situ. (After Brace, W. F., 1984, Permeability of crystalline rocks: new in situ measurements. *J. Geophys. Res.* 89:4327–30.)

When the permeability is very low, a transient pressure pulse method can be utilized. In this method a fluid pressure at the source is "instantaneously" increased by Δp. Assuming the system geometry is known, the permeability can be computed from the time it takes for this transient pressure pulse to reach the recording device (Brace et al., 1968).

Variation with Pressure, Temperature, and Deviatoric Stress

In situ, rocks are exposed to a wide variety of pressure (lithostatic pressure P and fluid pressure p), temperature, and deviatoric stress conditions. These variable conditions can result in major changes in rock permeability.

Increasing lithostatic pressure reduces the permeability by progressively closing pores (fig. V.5). Because fractures close easily when pressure is applied (equ. IV.14), pressure effects are very pronounced in rocks where permeability is primarily due to fractures (fig. V.6). For isotropic pores, as in sandstones or other granular materials, pressure effects are very small. In the intermediate case where both pores and fractures coexist, the effects can be large or small depending on whether the fractures are required to maintain the interconnectivity of the pore network (fig. V.7). For example, the porosity at grain contacts in

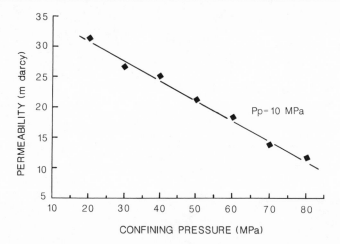

Fig. V.5. Variation of k with P (Fontainebleau sandstone, $\phi = 12\%$) for the case of isotropic pores. (After David, C., and M. Darot, 1989, Permeability and conductivity of sandstones. In *Rocks at Great Depth*, ed. V. Maury and D. Fourmaintraux, Balkema, pp. 203–120.)

sandstones can be viewed as a fracture porosity. The comparison of experimental results and models show that the closing of fractures is controlled by elastic processes (as in equ. IV.14) up to a certain value of pressure. Above this value, the surface roughness controls further closure (fig. V.8). Thus there are two ways that a non-negligible permeability can be preserved at elevated pressures: the first is to have an interconnected porosity composed of "isotropic" pores, the second is to have fractures with rough surfaces. In the second case, interconnection is more likely to be assured.

Pore pressure p has the opposite effect to lithostatic pressure P (fig. V.9): increasing p tends to open the pores. The two effects of p and P do not in general compensate for each other. That is, the differential pressure $P_D = P - p$ is not the effective pressure. As discussed by Bernabe (1987), we can define the effective stress P_{eff} as the lithostatic pressure that applied by itself would give the same value of k:

$$k(P_{\text{eff}}, p = 0) = k(P, p).$$
$\hspace{6cm}$ V.7

If $k(P, p)$ is a single-valued function, this definition can be in general approximated by the relation

$$P_{\text{eff}} = P - np$$
$\hspace{6cm}$ V.8

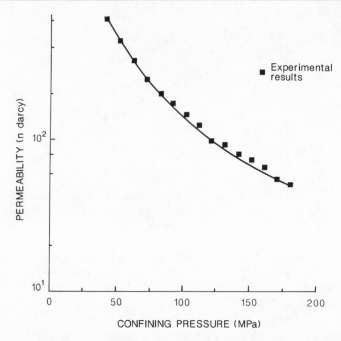

Fig. V.6. Variation of k with P for a medium with fracture porosity. (After Bernabe Y., 1986, The effective pressure law for permeability in Chelmsford Granite and Barre Granite. *Int. J. Rock Mech. Min. Sci. and Geomech. Abst.* 23: 267–25 and Gavrilenko, P., and Y. Guéguen, 1989, Pressure dependence of permeability: a model for cracked rocks, *Geophys. J. Int.* 98:159–72.)

where the coefficient n is not necessarily constant but varies with the pressure range (P, p). When the sample is submitted to pressure (P, p) cycling, one generally observes an irreversible variation of k due to irreversible changes in pore structure. This is a serious problem faced when one wishes to extrapolate laboratory data to in situ conditions.

Temperature affects the permeability through two distinct processes: (1) thermally inducing microfractures, and (2) enhancing dissolution-recrystallization reactions. Thermally induced microfractures result from the differential dilation of crystals, due to either a strong temperature gradient or to large differences in the mineral thermal expansion coefficient. The resulting permeability increase will be a function of the density and geometry of induced microcracks. Dissolution-recrystallization processes are controlled by temperature, pressure, and deviatoric stress gradients. At the large scale, the effects of temperature and pressure can result in fluid circulation and the transfer of mass from

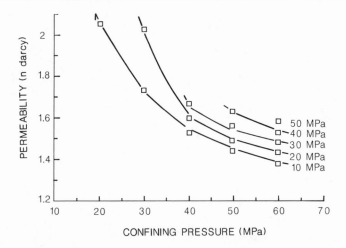

Fig. V.7. Variation of k with P for the intermediate case (Fontainebleau sandstone, $\phi = 6\%$, isotropic pores plus crack grain porosity). (After David, C., and M. Darot, 1989, Permeability and conductivity of sandstones. In *Rocks at Great Depth*, ed. V. Maury and D. Fourmaintraux, Balkema, pp. 203–10.)

zones of elevated solubility to zones of reduced solubility. At the local scale, the processes of dissolution-recrystallization are driven by "stress concentrations" at grain contacts and result in progressive rock compaction and cementation. The suite of complex temperature effects cannot be adequately addressed in a general manner and are best illustrated by the particular situation at hand.

Deviatoric stresses have already been mentioned in relation to dissolution-recrystallization reactions. These processes depend on the pressure of confinement, fluid pressure, and the deviatoric stresses. They result in a reduction in porosity and permeability by kinetic processes, the details of which are not well understood. Deviatoric stresses can also have the opposite effect, producing microcracks with an accompanying increase in permeability.

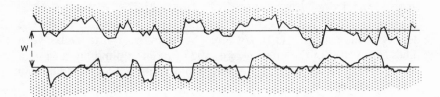

Fig. V.8. Modeling the surface roughness of a fracture.

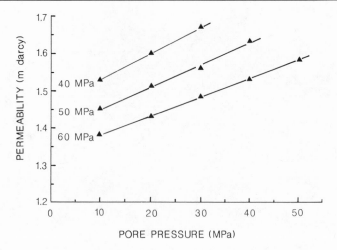

Fig. V.9. Variation of k with pore pressure (same sample as fig. V.7). (After David, C., and M. Darot, 1989, Permeability and conductivity of sandstones. In *Rocks at Great Depth*, ed. V. Maury and D. Fourmaintraux, Balkema, pp. 203–10.)

PERMEABILITY MODELS

According to Scheidegger (1974), one can group the permeability models into two families: models involving networks of capillary tubes and models utilizing the notion of hydraulic radius. In the first case one attempts to describe the permeability in terms of capillaries with a distribution of radii, and thus the model is an approximation to the real system (Guéguen and Dienes, 1989). In the second case one introduces the notion of an equivalent porous medium (Walsh and Brace, 1984). Dullien (1979) calls the first class "statistical" models and the second "geometrical."

Tube and Crack Models

These models are part of the first group cited. The pore volume of a rock is modeled as a homogeneous and isotropic distribution of tubes (capillaries) which have a variable length d and a variable radius r. The density of tubes is $N = (1/\bar{l}^3)$, where $\bar{l}$ is the average distance between tubes (fig. V.10). In the interior of a tube of radius r, fluid flows with an average velocity $\bar{v}$ which is given by Poiseuille's law (Landau and Lifshitz, *Fluid Mechanics*):

$$\bar{v} = -\frac{r^2}{8\eta}\frac{dp}{dx}.$$

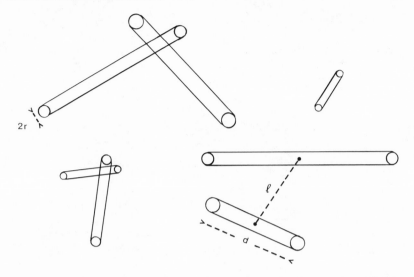

Fig. V.10. Capillary tube model: r is the radius, d the length, and l the average distance between two tubes.

Assuming that the radii do not vary greatly and combining with V.2, we find $q = -\dfrac{\bar{r}^2}{8\eta}\dfrac{dp}{dx}\phi$. Comparing this result with V.1, one obtains:

$$k = \frac{\bar{r}^2}{8}\phi. \qquad\qquad \text{V.9}$$

Using the expression $\phi = \dfrac{\pi \bar{r}^2 \bar{d}}{\bar{l}^3}$ in conjunction with the approximation $\overline{(r^n)} = (\bar{r})^n$ we find:

$$k = \frac{\pi}{8}\frac{\bar{r}^4 \bar{d}}{\bar{l}^3}. \qquad\qquad \text{V.10}$$

The permeability is controlled by the distributions of the three microvariables d, r, and l.

In the same way, one can calculate the permeability resulting from a distribution of cracks of aperture $2w$ and diameter $2c$ (fig. V.11). Utilizing V.2 and the computation for the average flow velocity $\bar{v}$ between two parallel planes (Landau and Lifshitz, *Fluid Mechanics*), it can be shown:

$$\bar{v} = -\frac{\bar{w}^2}{3\eta}\frac{dp}{dx}, \qquad q = \bar{v}\phi.$$

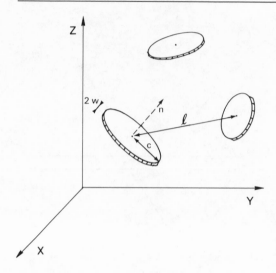

Fig. V.11. Crack model: w is the half-width of the aperture, c the radius, and l the average distance between two cracks.

Thus by comparison with V.1, we find:

$$k = \frac{\overline{w}^2}{3}\phi.$$

V.11

With same approximations discussed previously, $\phi = 2\pi\dfrac{\overline{c}^2\overline{w}}{\overline{l}^3}$, and

$$k = \frac{2\pi\overline{c}^2\overline{w}^3}{3\overline{l}^3}.$$

V.12

Here k is again expressed in terms of three microstructural parameters, the microvariables c, w, and l.

Percolation Effects

The preceding models assume that there exists a perfect interconnection between the tubes or cracks, which is not always the case. As we have discussed in chapter III, percolation theory evaluates the degree of network interconnectivity. The results V.9 and V.12 must therefore be multiplied by a factor f, which represents the fraction of capillaries or cracks that are attached to the percolating cluster (are interconnected). Thus for example, V.9 should be

$$k = \frac{\pi}{8}f\overline{r}^2\phi.$$

V.13

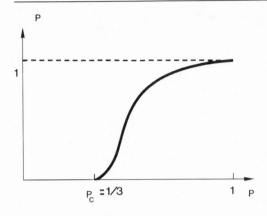

Fig. V.12. Variation of k is a function of the probability p: p_c is the percolation threshold.

Near the percolation threshold

$$f \sim (p - p_c)^2$$

where p is the probability that two capillaries or two cracks are interconnected. For the case of cracks, the excluded volume defined in chapter III is $V_{ex} = \pi^2 \bar{c}^3$. Using the results of chapter III (equ. III.12),

$$p \approx \frac{\pi^2 \bar{c}^3}{4 \bar{l}^3} \, .$$

For the case of capillary tubes, the excluded volume is $V_{ex} = 2\bar{d}^2 \bar{r}$, thus

$$p \approx \frac{\bar{d}^2 \bar{r}}{2 \bar{l}^3} \, .$$

Below the percolation threshold defined by p_c ($p < p_c = 1/3$ for the model discussed in chapter III), the interconnections are finite and $k = 0$. The permeability increases very rapidly above p_c and approaches the values given by V.9 and V.11 (fig. V.12).

Hydraulic Radius Models

The dynamic radius m is defined by the ratio (pore volume)/(pore surface): $m = \dfrac{V_p}{S_p}$. For example, if the porosity is due to cylindrical capillaries, $V_p = \dfrac{\pi \bar{r}^2 \bar{d}}{\bar{l}^3}$ and $S_p = \dfrac{2\pi \bar{r} \bar{d}}{\bar{l}^3}$, thus $m = \dfrac{\bar{r}}{2}$. From the previous considerations we have seen that for capillaries of radius m, $k \sim m^2 \phi$ (equ. V.9). If we adopt a model of an equivalent medium where the radius of capillaries is m, then

$$k = am^2 \phi = \frac{a}{S_o^2} \frac{\phi^3}{(1 - \phi)^2}$$

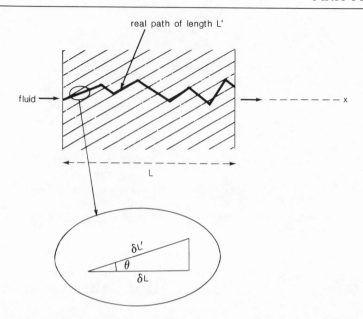

real path of length L'

fluid →

x

L

δL'

θ

δL

Fig. V.13. Notion of tortuosity: L' is the real flow path of the fluid and L is the apparent path. $\tau = \dfrac{\delta L'}{\delta L} = \dfrac{L'}{L}$.

where a is a dimensionless constant of magnitude close to unity and S_o is the surface area of the pores per unit volume of solids (S_o = specific surface area). In effect $\phi = V_p/V$ and $S_p/V = S_o(1 - \phi)$ for a volume of rock V.

The preceding relation is called the *Carman-Kozeny equation*. It can be modified to incorporate the effects of percolation in a form different than previously discussed. First let us introduce the notion of tortuosity. Tortuosity τ is defined through the ratio $\tau = L'/L$, where L' is the real length of the path through which the fluid flows and L is the apparent path length (fig. V.13). The local velocity of fluid is:

$$v_x = -\frac{r^2}{8\eta}\left(\frac{dp}{ds}\right)\cos\theta$$

where $\left(\dfrac{dp}{ds}\right)$ is the pressure gradient along the real path L' and $\cos\theta = \dfrac{\delta L}{\delta L'} = \dfrac{L}{L'}$ by hypothesis.

$$v_x = -\frac{r^2}{8\eta}\left(\frac{dp}{ds}\right)\frac{L}{L'} \quad \text{and} \quad \frac{ds}{dx} = \frac{\delta L'}{\delta L} = \tau.$$

Thus

$$v_x = -\frac{r^2}{8\eta}\left(\frac{dp}{dx}\right)\left(\frac{L}{L'}\right)^2.$$

Comparing with Darcy's law, one obtains:

$$k = \frac{1}{b\tau^2}m^2\phi$$

where the constant a has been replaced by a new constant $\frac{1}{b\tau^2}$.

Comparing this result and V.13, one can associate $f \sim \frac{1}{\tau^2}$. When the tortuosity becomes infinite, it is no longer connected and $f \to 0$. The notion of tortuosity and of percolation are directly related.

Permeability-Porosity Relations and Evolution Laws

Although it has often been remarked that "there exists no simple satisfactory relationship between porosity and permeability" (Scheidegger, 1974), the search for such correlations continues even today. This can be explained by the fact that under certain restrictive conditions, such relations have been developed and verified by their usefulness. The empirical constants that are introduced can be adjusted to make the correlations satisfactory for the data set used. But it is important to keep in mind that in fact the permeability is not determined by the porosity ϕ, but by the porosity microstructure.

The capillary tube and crack models examined previously illustrate this point quite clearly. Using these models, an exact calculation (Guéguen and Dienes, 1989) yields the results shown in table V.1. The factor f is the percolation factor ($f = 0$ if $p < p_c$, $f \sim (p - p_c)^2$ if $p \approx p_c$, $f = 1$ if $p \gg p_c$). It depends on the three introduced microvariables. The two quantities k and ϕ are also obtained as functions of the three microvariables $\bar{d}, \bar{r}, \bar{l}$ or $\bar{w}, \bar{c}, \bar{l}$. There exists a unique relation

Table V.1. PERMEABILITY AND POROSITY FOR TWO MODELS

Model	k	ϕ
Capillary	$\dfrac{\pi}{32}f\dfrac{\bar{d}\bar{r}^4}{\bar{l}^3}$	$\pi\dfrac{\bar{r}^2\bar{d}}{\bar{l}^3}$
Crack	$\dfrac{4\pi}{15}f\dfrac{\bar{w}^3\bar{c}^2}{\bar{l}^3}$	$2\pi\dfrac{\bar{c}^2\bar{w}}{\bar{l}^3}$

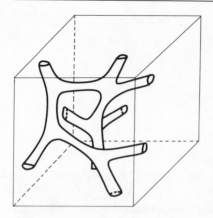

Fig. V.14. Idealized schematic for the porosity structure of a sandstone: the pores are localized by the grain packing and can be viewed as capillaries of variable cross-section.

between k and ϕ only if one of the three microvariables is fixed, or if there exists an imposed relation between two of the microvariables. Equations V.9 and V.11 are examples of such relations between k and ϕ: in reality these relations also depend on the variables $\bar{r}$ and $\bar{w}$. For the case where $\bar{r}$ and $\bar{w}$ are constant, V.9 and V.11 would effectively provide permeability-porosity relationships.

Many other cases of interest can be considered. Suppose for example that the density of pores varies but that the characteristics of the pores, tubes, or cracks, remain constant (d and r are constant, or w and c are constant). If the density is sufficiently high for f to be close to 1, one obtains relations of the type shown in V.9, with $k \sim \phi$. If on the other hand the density is constant but the radius r varies (case of tubes) or w (case of cracks) one obtains $k \sim \phi^2$ (tubes) or $k \sim \phi^3$ (cracks).

Correlations between k and ϕ for sedimentary rocks are primarily determined by the rock history which alters the porosity structure. For example if the porosity is composed of capillaries, the density of capillaries ($\bar{l}$), the radius ($\bar{r}$), and their length ($\bar{d}$) are all changing with time (t) during the rock history:

$$\bar{l} = \bar{l}(t), \quad \bar{d} = \bar{d}(t), \quad \bar{r} = \bar{r}(t).$$

If we consider a sand which is composed of approximately equal size quartz grains, the pores are defined by the contact points of nearest neighbor grains (fig. V.14). As a first approximation $\bar{l} = \bar{d} = \bar{l}_g$ = average grain diameter. Thus

$$k = \frac{\pi}{32} f \frac{\bar{r}^4(t)}{\left(l_g^2\right)}, \quad \phi = \frac{\pi \bar{r}^2(t)}{\left(l_g^2\right)}, \quad \Rightarrow k \approx \phi^2.$$

Other more complex models, which describe the progressive cementation resulting from dissolution-recrystallization reactions, lead to different relations between k and ϕ.

For the case of fractured rocks, the variation of permeability with pressure is a particularly important problem. From table V.1 and the results of chapter IV, it can be seen that the evolution of k and ϕ are essentially controlled by $\bar{w}$, if $\bar{c}$ and $\bar{l}$ remain constant. At a pressure $P = A\dfrac{E}{2(1 - \nu^2)}$ (equ. IV.15), the fractures of aspect ratio $A = \dfrac{w}{c}$ are closed. The total closure of fractures leads to a decrease in the fracture density and an increase in $\bar{l}$. In this way one expects to see the effects of percolation and a vanishing k. However, this is not what is observed in nature (fig. V.6). Due to the roughness of fracture surfaces (fig. V.8), the elastic closure law is not applicable when the asperities of the fracture surfaces come into contact. Due to the increasing contact between asperities, the aperture w decreases much more slowly with increasing P than what is predicted by the elastic law. The fracture closure model ignores surface roughness and therefore is only useful when the confining pressures are relatively small. A more accurate model is possible but requires introduction of additional (statistical) parameters characterizing the surface roughness (fig. V.8).

PROBLEMS

V.a. Converting Permeability to Hydraulic Conductivity

Darcy's law takes the form $q = -K\dfrac{d}{dz}\left(\dfrac{p}{\rho g} - z\right)$ in the presence of a gravitational field; g is the acceleration of gravity ($g = 9.81$ m sec^{-2}) and ρ is the density of the fluid.

(1) Calculate the hydraulic conductivity K in (m sec^{-1}) as a function of the permeability k.
(2) Evaluate K numerically for $k = 1$ D, the saturating fluid being water.

V.b. Deviations from Darcy's Law and the Mean Free Path of Gas Molecules

Darcy's law is not applicable for the flow of gases at very low pressures (the Klinkenberg effect). In that case Darcy's law becomes $q = -\dfrac{k_a}{\eta}\dfrac{dp}{dz}$, where $k_a = \left(b + \dfrac{b'}{p}\right)$; b and b' being two constants. This effect results from the existence of non-zero velocities at the boundaries of the capillaries, and occurs when the mean free path λ of the molecules becomes comparable to the pore radius.

Define (discuss) the domain of (p, r) such that $\lambda \geq r$. Assume that the gas is perfect. Utilize $\frac{1}{2}mv^2 = \frac{3}{2}kT$ and $\nu = $ number of collisions/ second with the other molecules $= 16 \ a^2 p \sqrt{\dfrac{\pi}{mkT}}$. T is the temperature, m the mass, and a the radius of the molecules. $k = 1.38 \times 10^{-23} JK^{-1}$.

V.c. Variation of Permeability with Pressure

Utilizing the results of chapter IV (equs. IV.14 and 15 and table V.1), determine how the permeability varies with pressure. Assume that the fractures have the shape of disks, that the closure of the fractures is elastic, and that the fractures are all connected.

V.d. Percolation Threshold

(1) Calculate the value of l (average distance between fractures) at the percolation threshold. Assume that the fractures are disk shaped and of radius $c = 200 \ \mu$m.

(2) Calculate the maximum value of the permeability for a population of identical fractures with radius $c = 200 \ \mu$m and aperture $w = 1 \ \mu$m. (Utilize the results of chap. III.)

REFERENCES

Bernabe, Y. 1987. The effective pressure law for permeability during pore pressure and confining pressure cycling of several crystalline rocks, *J. Geophys. Res.* 92:649–57.

Brace, W. F., J. B. Walsh, and W. T. Frangos. 1968. Permeability of granite under high pressure, *J. Geophys. Res.* 73:2225–36.

Dullien, F.A.C. 1979. *Porous Media, Fluid Transport and Pore Structure.* New York: Academic Press.

Guéguen, Y., and J. Dienes. 1989. Transport properties of rocks from statistics and percolation. *Mathematical Geology* 21:1–13.

Landau, L., and E. Lifshitz. 1971. *Fluid Mechanics.* Moscow: Editions Mir.

Scheidegger, A. E. 1974. *The Physics of Flow Through Porous Media.* Toronto: University of Toronto Press.

Walsh, J. B., and W. F. Brace. 1984. The effect of pressure on porosity and the transport properties of rock. *J. Geophys. Res.* 89:9425–31.

VI. Mechanical Behavior of
Fluid-Saturated Rocks

TEMPERATURE, pressure, and stress determine the mechanical behavior of rocks. In chapter IV we examined the three main regimes of rock deformation: elasticity, plasticity, and fracture. The presence of a fluid phase within a rock can modify each of these responses. In particular, the equilibrium conditions and kinetics can be strongly modified.

The existence of fluids in the crust has been proven down to at least 10 km depth. The importance of fluid-rock interactions in the crust is well recognized, through their control of many of the geological, geochemical, and geophysical phenomena such as diagenesis, ore deposits, earthquakes, and heat flow. Due to the presence of fluid, the mechanical behavior of the crust is often modeled as that of a porous, fluid-saturated medium.

LINEAR POROELASTICITY

To describe small deformations of a saturated porous medium, the theory of elasticity must be generalized to poroelasticity. Poroelastic theory describes the mechanical behavior at moderate pressure and temperature conditions. The new feature of poroelasticity is that it takes into account the fluid phase. Two additional parameters are required in order to describe the thermodynamic state of the fluid: its pressure and its volume (or mass). The isothermal theory of poroelasticity was first presented by Biot (1941) and later reformulated by Rice and Cleary (1976). In the same way that stress-strain relations are linear in elasticity, stress-strain relations (including fluid pressure-strain relations) are assumed to be linear in poroelasticity. Non-isothermal effects will not be considered in this chapter. Mathematically they can be incorporated through an additional term which describes linear changes in strains due to a change in temperature (Palciauskas and Domenico, 1982; McTigue, 1986). Here we shall follow the formulation of Rice and Cleary (1976).

Concept of Effective Pressure

The concept of effective pressure is fundamental to poroelasticity. Let P denote the pressure as defined in chapter IV (average of the stress tensor) and p be the fluid pressure, that is, the equilibrium fluid pressure inside the connected pores. In addition, let v be the fluid volume per unit volume of rock and m the fluid mass per unit rock volume. Then $v = m/\rho$, where ρ is the fluid density. For a fully saturated porous medium, $v = \phi$. In general the fluid is compressible and the bulk modulus K_f and density ρ depend on p.

The effective pressure P^* is defined through the linear equation:

$$P^* = P - \alpha p \qquad\qquad \text{VI.1}$$

where α is a coefficient which will be computed below. The effective pressure is not the differential pressure P_D, which is simply the difference between P and p:

$$P_D = P - p. \qquad\qquad \text{VI.2}$$

The effective and differential pressures are identical only for the particular case $\alpha = 1$. Equation VI.1 is formally similar to equation V.8 which was introduced in chapter V. There is however a fundamental difference between both definitions, that of chapter V and the one above. The effective pressure was introduced in chapter V through an operational definition. Thus one could define N different effective pressures using N equivalent operational definitions: one for permeability, another one for conductivity, etc.

Definition VI.1, when completed by the calculation of α, is unique and is a constitutive relation of linear poroelasticity. Coefficient α is:

$$\alpha = 1 - \frac{K}{K_s} \qquad\qquad \text{VI.3}$$

where K_s is the bulk modulus of the solid part of the medium (rock without pores), and K is the bulk modulus of the "skeleton," that is, the rock with pores (frame modulus). Equation VI.3 can be derived from an analysis proposed by Nur and Byerlee (1971): the deformation of a solid submitted to both a hydrostatic pressure P and a pore pressure p can be obtained by superimposing two states of equilibrium (fig. VI.1). The first one corresponds to a pressure $(P - p)$ applied to the external surface of a rock (outer envelope) which has an apparent bulk modulus K. K is the "frame" modulus (effective modulus of the porous rock with empty pores). It follows that the equilibrium state for figure VI.1 (a) corresponds to a volumetric strain:

$$\left(\frac{\delta V}{V}\right)^a = (\varepsilon_{kk})^a = -\frac{P-p}{K} \quad (a).$$

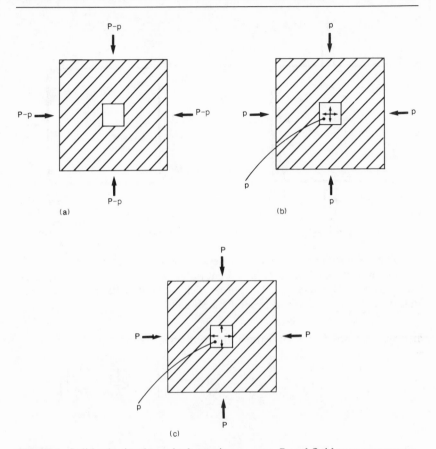

Fig. VI.1. Solid submitted to a hydrostatic pressure P and fluid pore pressure p. State (c) is obtained by superimposing state (a) where pressure $(P - p)$ is applied on the external surface, and state (b) where pressure p is applied on both the external and internal surfaces.

The equilibrium state of configuration (b) is that of a solid submitted to hydrostatic pressure p on the external surface of the solid and internal surfaces of pores. In this case, the pores can be ignored because they are at the same pressure as the solid phase. Because the modulus of the solid without pores is K_s, it follows that:

$$\left(\frac{\delta V}{V}\right)^b = (\varepsilon_{kk})^b = -\frac{p}{K_s} \quad (b).$$

Because we are using a linear theory, we can superimpose states (a) and (b) to get state (c), the state where the solid is submitted to a

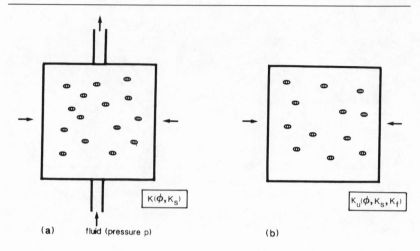

Fig. VI.2. Drained and undrained regimes. (*a*) Drained regime: fluid pressure *p* is kept constant. Fluid can flow in and out of the pores. (*b*) Undrained regime: fluid mass *m* is kept constant. The porous rock does not exchange fluid with the outside and can be described by an effective medium.

hydrostatic pressure P and the fluid to a pressure p:

$$\left(\frac{\delta V}{V}\right) = \varepsilon_{kk} = -\frac{P-p}{K} - \frac{p}{K_s} = -\frac{1}{K}\left[P - p\left(1 - \frac{K}{K_s}\right)\right]$$

or:

$$\varepsilon_{kk} = -\frac{P - \alpha p}{K} = -\frac{P^*}{K} \qquad\qquad \text{VI.4}$$

where $\alpha = 1 - (K/K_s)$, in agreement with VI.3. The effective pressure P^* plays the same role as P, the hydrostatic pressure in chapter IV.

Drained and Undrained Deformations

The existence of a fluid phase implies that there are two distinct deformation regimes to be considered: the drained and undrained regimes. Here we shall restrict ourselves only to the isothermal case. In the drained regime, the fluid pressure remains constant while the amount of fluid within the rock changes. Recall that the porosity is assumed to be completely connected so that the fluid pressure is uniform. In the undrained regime, fluid mass remains constant and the fluid pressure is variable (fig. VI.2).

Stress-strain relations in the *drained regime* are the same as in ordinary elasticity, modified only by the use of effective pressure P^*

instead of P. More precisely, equations IV.6 remain valid if one substitutes effective stresses σ_{ij}^* for the stresses σ_{ij}:

$$\sigma_{ij}^* = \sigma_{ij} + \alpha p \delta_{ij}. \qquad \text{VI.5}$$

Shear stresses (off diagonal terms) remain unchanged. One then finds the following expressions for the hydrostatic and shear deformations:

(a) for hydrostatic compression: $\varepsilon_{kk} = \dfrac{\sigma_{kk}^*}{3K}$

(b) for shear deformation $(i \neq j)$: $\varepsilon_{ij} = \dfrac{\sigma_{ij}^*}{2\mu}$. $\qquad \text{VI.6}$

Equations VI.6 are the basic equations of poroelasticity for the drained regime and replace equations IV.6 of ordinary elasticity. The theory for the drained regime contains three poroelastic constants, K, μ, and α, which characterize the medium. Note that the existence of a fluid at pressure p modifies the stress-strain relations only for hydrostatic compression.

In the *undrained regime*, pressure p is no longer an independent, externally controlled parameter. In this case the thermodynamic parameter which is externally controlled is v (fluid volume per unit volume of rock) or m (fluid mass per unit volume of rock, $m = \rho v$), and p takes on a value which depends on the internal deformation state of the medium. Thermodynamically, the rock-fluid system is a closed one and can be described through equations VI.6. However such a description is not very useful because p and σ_{ij}^* are unknown and depend on the complex pore microstructure. Another approach is to consider the closed rock-fluid system as an equivalent effective medium, with elastic constants K_u and μ_u (index u reflecting undrained conditions). This second description assumes that the fluid phase is the second phase considered in the models of chapter III. Since shear strains are unchanged (as shown by VI.5), $\mu_u = \mu$. The bulk modulus K_u is a new constant, to be added to the three previous ones. Poroelasticity thus becomes a theory with four independent constants (instead of two in elasticity): K, μ, α, K_u. Using the effective medium approximation, equations VI.6 in the undrained regime become

(a) hydrostatic compression regime: $\varepsilon_{kk} = \dfrac{\sigma_{kk}}{3K_u}$

(b) shear regime $(i \neq j)$: $\varepsilon_{ij} = \dfrac{\sigma_{ij}}{2\mu}$. $\qquad \text{VI.7}$

One can also use the Poisson coefficient for the undrained regime $\nu_u = \dfrac{3K_u - 2\mu}{2(3K_u + \mu)}$ (see table IV.2).

The concept of effective pressure, although still valid, is not very useful in the undrained regime, because p is not an independent parameter. Using VI.4, one finds:

$$p = \frac{1}{\alpha}[P + K\varepsilon_{kk}].$$

Because $P = -\dfrac{\sigma_{kk}}{3} = -K_u\varepsilon_{kk}$ (equ. VI.7), one finds that in the undrained regime, the fluid pressure change p_u in response to a volumetric strain ε_{kk} is:

$$p_u = -\left(\frac{K_u - K}{\alpha}\right)\varepsilon_{kk}. \qquad\qquad \text{VI.8}$$

By combining equations VI.7 and 8 and introducing the dimensionless coefficient $B = \dfrac{1}{K_u}\left(\dfrac{K_u - K}{\alpha}\right)$, the fluid pressure increase p_u can be expressed in terms of the increase in confining pressure $-\dfrac{\sigma_{kk}}{3}$ for undrained conditions.

$$p_u = -B\frac{\sigma_{kk}}{3}$$

$B = \dfrac{1}{K_u}\left(\dfrac{K_u - K}{\alpha}\right)$ is Skempton's coefficient and is one of the most often measured rock parameters.

Variation of the Fluid Mass as a Function of Deformation

Equations VI.6 and 7 are constitutive relations of poroelasticity which describe how the deformations and stresses are related. However, they do not give any information on the fluid mass variations within the sample. Changes in m also depend on two parameters: the strain of the porous rock (ε_{kk}) and the fluid pressure p. To determine $m(\varepsilon_{kk}, p)$, let us consider the more general framework of thermodynamics and choose ε_{ij}, T, and p as the independent thermodynamic variables. Denote the internal energy per unit volume of rock as $u(\varepsilon_{ij}, v, s)$ and the free energy as $f(\varepsilon_{ij}, p, T) = u - Ts - pv$. For isothermal deformations, infinitesimal variations $d\varepsilon_{ij}$ and dp result in a variation df:

$$df = \sigma_{ij}\, d\varepsilon_{ij} - v\, dp.$$

Because df is a total differential, one finds:

$$\left(\frac{\partial \sigma_{ij}}{\partial p}\right)_{\varepsilon_{ij}} = -\left(\frac{\partial v}{\partial \varepsilon_{ij}}\right)_p = -\frac{1}{\rho}\left(\frac{\partial m}{\partial \varepsilon_{ij}}\right)_p.$$

From equation VI.6, one derives:

$$\left(\frac{\partial \sigma_{ij}}{\partial p}\right)_{\varepsilon_{ij}} = -\alpha \delta_{ij} \quad \text{and} \quad -\frac{1}{\rho}\left(\frac{\partial m}{\partial \varepsilon_{ij}}\right)_p = -\alpha \delta_{ij}.$$

Assuming that all changes are very small, the change in the fluid mass content can be expressed as $\Delta m = m - m_o = \alpha \rho \varepsilon_{kk} + Ap$. The constant A is obtained from the condition that $\Delta m = 0$ when $p = p_u$:

$$\Delta m = \alpha \rho \left[\varepsilon_{kk} + \frac{\alpha p}{K_u - K}\right]. \qquad \text{VI.9}$$

Equation VI.9 expresses how the fluid mass content varies within the rock and complements equations VI.6 and 7. Because we are using linear approximations to the constituent equations, the density ρ has a constant value ρ_o in equation VI.9. In reality $\rho = \rho(p)$, but the pressure dependence of ρ would introduce higher-order terms in VI.9.

The fluid volume change Δv is derived from VI.9 using

$$\Delta m = \rho \Delta v + v \Delta \rho$$

where $\Delta \rho = \rho_o\left(\dfrac{p}{K_f}\right)$ is the change in ρ and K_f is the fluid bulk modulus. Note that for a fully saturated medium $v = \phi$. Thus

$$\Delta v = \alpha \varepsilon_{kk} + \left(\frac{\alpha^2}{K_u - K} - \frac{\phi}{K_f}\right)p. \qquad \text{VI.10}$$

The pressure dependence of ρ introduces a first-order term in VI.10, whereas it introduced a second-order term for Δm in VI.9. This last equation is often written as:

$$p = -\alpha M \varepsilon_{kk} + M \Delta v$$

where the modulus M is defined as: $\dfrac{1}{M} = \dfrac{\alpha^2}{K_u - K} - \dfrac{\phi}{K_f}.$

Interpretation of Poroelastic Constants

Linear poroelastic theory uses four independent constants: K, μ, K_u, and α. Other choices are also possible: ν_u can be used instead of K_u, B instead of K, etc. Modulus K is the frame modulus (or skeleton modulus), that is, the effective bulk modulus of a saturated porous rock at constant pressure. K is obtained by calculating the effective modulus of the rock with empty pores ($p = 0$). Using the methods presented in chapter III, one finds that K depends on K_s and ϕ. Modulus K_u is the effective bulk modulus of a porous rock which contains a second phase (the fluid) of compressibility K_f. From the theory presented in chapter III, it is possible to derive K_u as a function of K_f, K_s, and ϕ. In general, $K_u > K$ because when the fluid cannot flow out of the rock, the rock is stiffer and deforms less.

Without explicitly deriving K and K_u, we can also obtain a general relation between both of these moduli from the previous equations. In the special case when $p \equiv P$, the solid medium is submitted to a homogeneous pressure. From the definition of K_s, this implies that $P = -K_s \varepsilon_{kk}$ and $p = -K_s \varepsilon_{kk}$. In this case $\dfrac{\Delta v}{v} = \varepsilon_{kk}$. Because the solid medium is submitted to a homogeneous pressure (and hence a homogeneous strain), the fluid can be ignored and the rock appears homogeneous, without pores, with a bulk modulus K_s. Using equation VI.10 and noting that $\phi = v$, one finds:

$$\Delta v = \phi \varepsilon_{kk} = \alpha \varepsilon_{kk} + \left(\frac{\alpha^2}{K_u - K} - \frac{\phi}{K_f} \right) (-K_s \varepsilon_{kk}).$$

Solving for K_u leads to the following equation:

$$K_u = K + \frac{\alpha^2}{\dfrac{\phi}{K_f} + \dfrac{(\alpha - \phi)}{K_s}}. \qquad \text{VI.11}$$

Equation VI.11 is independent of any effective medium calculation described in chapter III. Because $\alpha = 1 - \dfrac{K}{K_s}$, this equation is a relation between K_u and K which involves three other parameters: K_f, K_s, and ϕ. According to chapter III, K_u and K should depend not only on ϕ but also on the pore shapes. However, no geometrical factor appears explicitly in VI.11.

For an incompressible fluid, $K_f \to \infty$ and $K_u \to \left(K + \dfrac{\alpha^2 K_s}{\alpha - \phi} \right)$. For a very compressible fluid, $K_f \to 0$ and $K_u \to K$. Thus the undrained moduli can differ substantially for these two extreme situations. A practical occurrence of this case is when a rock is saturated by a liquid or a gas. Theory predicts a lower modulus for a gas. This result is

Table VI.1. NUMERICAL VALUES OF ν AND ν_u FOR THREE ROCKS

Rock	ν	ν_u
Westerley granite	0.27	0.30
Berea Sandstone	0.20	0.33
Clay soil	0.12	0.50

Source: Rice, J. R., and M. P. Clearly, 1976, Some basic stress diffusion solutions for fluid-saturated elastic porous media with compressible constituents. *Rev. Geophys. Space Phys.* 14:227–91.

important in oil/gas exploration and will be discussed further in chapter VII. Another interesting limit is that of a very porous rock, such as an unconsolidated sandstone. Then $K \ll K_s$, $\alpha \to 1$ and $\dfrac{1}{K_u} \to \dfrac{\phi}{K_f} + \dfrac{1 - \phi}{K_s}$, which is just equation III.5.

With the previous results, we can discuss how the dimensionless parameters B and ν_u vary:

$$B = \frac{K_u - K}{\alpha K_u} \quad \text{and} \quad \nu_u = \frac{3\nu + \alpha B(1 - 2\nu)}{3 - \alpha B(1 - 2\nu)}.$$

B varies between 0 and 1:

   $B \to 1$ when $K \to 0$ (porous, unconsolidated medium)
   $B \to 0$ when $K \to K_u$ (very compressible fluid).

ν_u varies between ν and $\dfrac{1}{2}$:

   $\nu_u \to \dfrac{1}{2}$ when both B and $\alpha \to 1$ (porous, unconsolidated medium)

   $\nu_u \to \nu$ when $B \to 0$ (very compressible fluid).

Table VI.1 gives examples of numerical values for ν and ν_u.

Fluid Diffusion

We assume that at every point within the rock, at the mini scale, local equilibrium has been established and that the quantities m and p are well defined. Equations VI.6, 7, and 9 then allow us to examine the general problem of fluid diffusion through porous rocks. When fluid flows, m and p vary from point to point at the marco scale. Fluid flow is described mathematically through Darcy's law (equ. V.1). In addition the conservation of mass equation implies that

$$\frac{\partial m}{\partial t} + \frac{\partial(\rho v_i)}{\partial x_i} = 0. \qquad \text{VI.12}$$

In the general case of fluid flow, p and m will also vary with time t. Within the linear approximation and small velocities v_i, the fluid density ρ will be taken as a constant. By combining equations VI.6, 9, and 12, Darcy's law and the equilibrium condition IV.12 (with zero body forces), one can derive the following equation:

$$\frac{\partial m}{\partial t} = c \nabla^2 m. \qquad \text{VI.13}$$

Here c is the hydraulic diffusivity:

$$c = \frac{1}{\alpha^2} \frac{k}{\eta} \left(\frac{K + \frac{4}{3}\mu}{K_u + \frac{4}{3}\mu} \right) (K_u - K). \qquad \text{VI.14}$$

The characteristic time constant for fluid to diffuse over a distance l is:

$$\tau \approx \frac{l^2}{c}.$$

Utilizing dimensionless parameters ν, B, and ν_u, VI.14 can also be expressed as:

$$c = \frac{2}{9} \mu \frac{k}{\eta} B^2 \left(\frac{1 - \nu}{1 - \nu_u} \right) \left(\frac{(1 + \nu_u)^2}{\nu_u - \nu} \right).$$

The water viscosity at room conditions is $\eta = 10^{-3}$ Pa s. Assuming that $\mu = 10^{11}$ Pa, $B = 1$, $\nu_u - \nu = 0.1$, $\nu = 0.25$, one finds

$$c = 0.1 \text{ m}^2 \text{ s}^{-1} \quad \text{if} \quad k = 10^{-3} \text{ Darcy.}$$

Fluid diffusion is a relatively slow process: $\tau \approx 30$ years, if $l = 10$ km.

Two extreme situations can be considered: that of a quasi-incompressible medium containing a compressible fluid, and that of a compressible medium containing an incompressible fluid. In the first case, variations of $m = \rho\phi$ result mainly from variations of ρ and VI.11 implies that $K_u \rightarrow K + \dfrac{K_f}{\phi}$ and $c \rightarrow \dfrac{k}{\eta} \dfrac{K_f}{\phi}$, when $\alpha \rightarrow 1$. In the second case, variations of m result mainly from variations of ϕ and $K_u \rightarrow \left(K + \dfrac{\alpha^2}{\alpha - \phi} K_s \right)$.

FRACTURE

When fluids are present, the conditions for crack stability are modified through two distinct processes of fluid-rock interactions. Interactions can be mechanical or chemical. In the first case fluid pressure can be sufficiently high and induce a stress field which leads to fracture. In the second case fluid pressures are not high enough for fracture, and chemical interactions control the kinetics of crack opening and closure.

Hydraulic Fracturing

Hydraulic fracturing occurs when a sufficiently high stress field is induced by fluid pressure. The role of the fluid is then only as an additional stress source. The geometry of the stress field depends on the geometry of the body considered and of the cavity in which fluid pressure is applied (cylindrical borehole, preexisting fracture, etc.). The process of fracturing is similar to what has been considered in chapter IV.

Hydraulic fracture is a technique used to measure in situ stresses around a borehole. This method assumes that a hydraulic fracture develops when the extensional stress at the borehole interface reaches the value of the tensile strength σ_1 of the rock. When the critical pressure $p = p_c$ is reached, the fracture propagates in the plane perpendicular to the least principal stress. The fluid pressure required to keep the fracture open is p_o. Pressure p_o is smaller than p_c and is equal to the stress component normal to the fracture plane (fig. VI.3). When the fracture is vertical, the three principal stresses are S_H, S_h, and S_v (compressive stresses are considered to be positive): $S_v = \rho g z$ and $S_h = p_o$. From a calculation of the maximum extensive stress at the borehole interface, one finds:

$$\sigma_1 = p_c - 3S_h + S_H. \qquad \text{VI.15}$$

Knowing σ_1, p_c, and $S_h = p_o$, equation VI.15 allows us to determine S_H. This equation is an approximation because we know that σ_1 is not an intrinsic rock property. The mechanical strength σ_1 depends on a, the maximum crack length within the rock (equ. IV.20). As we have seen in chapter IV, propagation of the hydraulic fracture results from the fact that a preexisting crack of length a has reached its stability limit. An exact derivation would yield a relation of the type:

$$K_I = f(S_h, S_H, R, a) = K_{IC}$$

where K_{IC} is the critical value of K_I. K_{IC} is considered to be an intrinsic property of a rock and depends on the elastic Young's modulus E and fracture surface energy (see chap. IV).

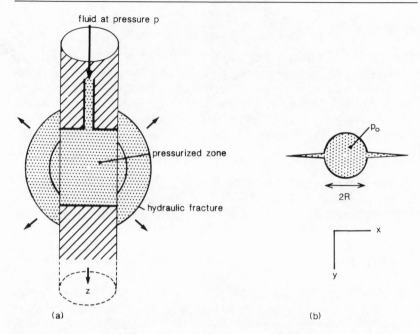

Fig. VI.3. Hydraulic fracture. (*a*) fracture occurs when the fluid pressure *p* inside the borehole reaches a critical value p_c. (*b*) the fracture is kept open when the value of *p* is in equilibrium with the minimum principal stress, i.e., $\sigma_{yy} = S_h = p_o$.

Laboratory results and in situ measurements have been compared by Rummel and Winter (1983) for various sandstones and granites. Such data allow the calculation of the critical crack lengths (table VI.2). Measured crack lengths in the laboratory are on the order of grain diameters. This result can be understood if one notes that potential cracks are often localized at grain boundaries. One can expect, however, a scale effect, that is, larger "in situ" crack lengths, and this is what is observed for sandstone.

Table VI.2. MAXIMUM CRACK LENGTHS AS INFERRED FROM LABORATORY AND IN SITU MEASUREMENTS

	K_{IC} $(MN\,m^{-3/2})$	a (mm), laboratory	a (mm), in situ
Sandstone	1.4–1.9	1.2	10
Granite	0.8–2.2	7	4

Source: F. Rummel and R. B. Winter, 1983, Fracture mechanics as applied to hydraulic fracturing stress measurements, *Earthquake Prediction Research* 2:33–45.

The previous discussion implies that it is the origin of the stress field which is characteristic of hydraulic fracture and not the mechanism of fracture. The mechanism is identical to that described in chapter IV.

Kinetics of Crack Opening and Closure

Below a critical threshold G_c or K_c, cracks submitted to stresses are observed to propagate slowly. As shown in figure VI.4, the presence of fluids modifies this behavior. Due to adsorption (see chap. II), the surface energy in a vacuum, $2\gamma_o$, is lowered to 2γ. For G values higher than 2γ, cracks propagate slowly: this is domain I shown on the plot of figure VI.4.

When v increases and reaches the value v_t, the crack velocity is greater than the fluid migration velocity. As a consequence, there will not be any fluid present at the crack tip and crack propagation then takes place as in a vacuum. This explains why the plot $v(G)$ exhibits a plateau (domain II) which ends at a point where the vacuum curve is intersected (domain III). The most important fluid is water (Atkinson, 1984). An explanation presented by Michalske and Freiman (1982) suggests that adsorption of water molecules leads to hydrolysis of Si-O bonds, which are then replaced by weak O-H bonds (fig. VI.5). For $G \gg 2\gamma$, equation VI.16 describes the propagation velocity reasonably well:

$$v = v_o(\gamma, T)\exp(\alpha G). \qquad \text{VI.16}$$

This law is not valid when $G \rightarrow 2\gamma$. In that limit $v \rightarrow 0$.

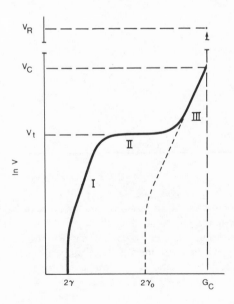

Fig. VI.4. Subcritical crack propagation: the dashed line corresponds to propagation in a vacuum, while the solid line corresponds to propagation in water.

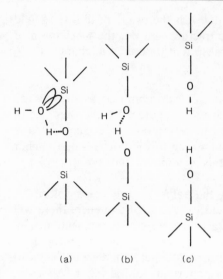

(a)   (b)   (c)

Fig. VI.5. Michalske and Freiman model: adsorption of a water molecule (*a*) leading to hydrolysis (*b*) and to rupture of the newly formed weak bonds.

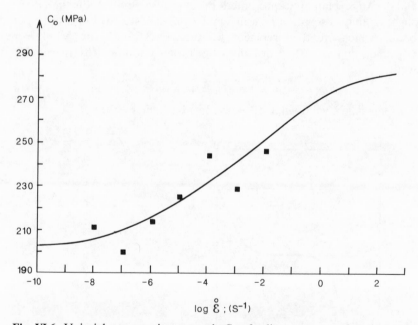

Fig. VI.6. Uniaxial compression strength C_o of a limestone as a function of strain rate $\dot{\varepsilon}$. Experimental data are noted by a solid square. The solid line is obtained by a theoretical calculation using single crack data. (From Reuschlé, T., M. Darot, and Y. Guéguen, 1989, Mechanical and transport properties of crustal rocks: from a single crack to crack statistics, *Phys. Earth and Planet. Int.* 55:353–60.

The propagation process is thermally activated, but the activation energy is low and cannot be measured with a high degree of accuracy. Kinetic laws such as $v = BK^n$ are also commonly used because they are convenient for mathematical calculations. However, they are not well supported on theoretical grounds. Applying equation VI.16 to a distribution of cracks, it is possible to predict how the strength in compression C_o depends on strain rate $\dot{\varepsilon}$. As shown in figure VI.6, the results agree reasonably well with theoretical predictions. The computations were performed with the assumption of a constant crack density, which is not true at very high stresses.

When $G < 2\gamma$, the crack front moves backward, which means that the crack closes. This process of "crack healing" is a complex process (Smith and Evans, 1984) which occurs through a series of different steps. First, there is a reorganization of cracks into a network of fluid inclusions of various shapes: pipes, spheres, etc. This reorganization implies that mass transport has been taking place through the fluid, along the fluid-crack interfaces. Once a crack has been transformed into a set of fluid inclusions, further changes can occur as a result of pressure solution, a process which will be examined in more detail later in this chapter. Kinetics of pressure solution depends on the degree of fluid saturation and on the temperature-pressure conditions. Smith and Evans' results show that, at 400°C and 200 MPa, crack healing in quartz occurs within a few days.

PLASTICITY

Crustal rocks can deform plastically at moderate pressure-temperature conditions when fluids are present. The specific mechanism of deformation is called pressure solution and is different from the ones described in chapter IV where only dry conditions were considered. The kinetics of pressure solution is sufficiently rapid so that large finite strains can be expected on geological time scales. This mechanism is of particular importance in the upper crust of the Earth.

Pressure Solution

Pressure solution is a deformation mechanism which transfers solid material through solution from highly stressed areas toward stress-free areas. The system progressively minimizes its total energy during this process. One can view pressure solution as a kind of diffusion creep (see chap. IV), where the diffusion takes place through the liquid phase. A sequence of three steps is observed: dissolution in regions where local stresses are high, mass transfer through solution, and recrystallization where local stresses are low. Pressure solution plays an important role

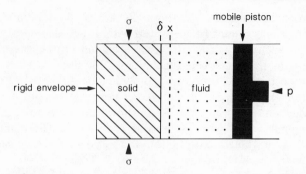

Fig. VI.7. A simple geometry to derive the chemical potential equation: a rigid cylinder contains a solid in contact with a fluid into which it can dissolve. The cylinder has an applied stress σ on its external faces, and the fluid pressure p is controlled through a mobile piston.

in many diagenetic processes (chap. II). Many other observed geological features suggest its importance: stylolites for instance, which are dissolution surfaces (Gratier, 1983).

One can view porosity evolution, in a fluid-saturated granular medium under an applied stress, in two complementary ways. The first view is to consider only the "solid" part of the medium. In a granular medium the highly stressed regions occur at the grain contacts, while the stress-free areas are the free-grain surfaces. Mass transfer is assumed to take place through the fluid film between grain contacts as a result of a chemical potential gradient. The chemical potential is given by Gibbs' equation:

$$\mu_f = f_s + p v_s \qquad\qquad \text{VI.17}$$

where f_s and v_s are the molar free energy and molar volume of the stressed solid, and p is the fluid pressure. The solubility is higher at grain-to-grain contacts where the normal stress is elevated. Due to this higher stress, the solution is locally above its normal value. Mass transfer through diffusion then takes place from highly stressed areas toward stress-free areas. Equation VI.17 is the basic relation which underlies all models which consider pressure solution. It expresses the equilibrium condition between a stressed solid and its solution.

To derive equation VI.17, consider the following system: a cylinder with cross-sectional area A contains a stressed solid which is in contact with a fluid into which it can dissolve (fig. VI.7). At any point on the fluid-solid interface, the equilibrium condition is that the chemical potential of the solid in solution μ_f must be equal to the chemical potential μ_s of the stressed solid. The solid is elastic, homogeneous, isotropic, and impervious, and the fluid-solid interface is planar. The applied pressure is kept constant through the use of a mobile piston and a stress σ which is applied to the cylinder by an appropriate loading system. The faces on which σ is applied are not in contact with the fluid phase. We also assume that the stressed solid is at equilibrium with the fluid phase and that the system (solid + fluid) cannot exchange matter with the external environment.

Let us assume that a reversible, infinitesimal perturbation is imposed on the equilibrium state. This perturbation is either from a volume fluctuation (displacement of the mobile piston) or from an entropy fluctuation (heat exchange). Such an infinitesimal perturbation at constant pressure and temperature produces dissolution or recrystallization of a thin layer δx of solid at the fluid-solid interface. The recrystallized solid has the same composition and is assumed to be under the same stress conditions as the initial solid phase. During the perturbation, the shape and area of the solid-fluid interface remain constant, that is, the interface is simply translated. If this was not true, an additional surface energy term should be added. Let ΔG be the free enthalpy change for the system. Then:

$$\Delta G = \Delta U + p\,\Delta V - T\,\Delta S.$$

At constant p and T, the equilibrium condition is $\Delta G = 0$. The internal energy U is assumed to be a function of the entropy S and volume V: $U = U(S,V)$, (Landau and Lifchitz, 1967). Let us consider the case where dissolution is occurring, that is, $\delta x < 0$. The opposite case, $\delta x > 0$, can be treated in the same manner. Because no mass exchange takes place between the system and the outside world, the total number of moles is kept constant:

$$A\,\delta x - v_s\,\delta n = 0$$

where v_s = molar volume of stressed solid, δn = number of moles of dissolved solid and A = area of the solid-fluid interface.

The volume change ΔV can be expressed as $\Delta V = (v_f - v_s)\delta n$, where v_f is the molar volume of the solid in solution. Similarly, the entropy change ΔS, which results from the heat exchange ΔQ, can be expressed as:

$$\Delta Q = T\,\Delta S = T(s_f - s_s)\delta n.$$

Here s_s = molar entropy of the stressed solid, s_f = molar entropy of solid in solution, and T = temperature. As δn moles are transferred from the solid to the fluid phase, the internal energy changes by ΔU:

$$\Delta U = (u_f - u_s)\delta n.$$

Here u_s = internal molar energy of the stressed solid and u_f = internal energy of the solid in solution. Because the piston moves in order to keep p constant, the amount of work ΔW done by the loading system is

$$\Delta W = -p\,\Delta V.$$

Note that the stress σ does no work because the wall of the cylinder is not allowed to move. The above analysis leads to:

$$\Delta G = (u_f - u_s)\delta n - T(s_f - s_s)\delta n + p(v_f - v_s)\delta n.$$

At equilibrium $\Delta G = 0$, thus

$$u_f - Ts_f + pv_f = u_s - Ts_s + pv_s. \qquad\qquad \text{VI.18}$$

The same result holds for crystallization ($\delta x < 0$). By definition, the chemical potential of the solid in solution μ_f is equal to its molar free enthalpy g_f (Landau and Lifchitz, 1967). This leads to:

$$\mu_f = g_f = u_f - Ts_f + pv_f.$$

But the molar free energy of the stressed solid is f_s:

$$f_s = u_s - Ts_s.$$

The above relations then lead to Gibbs' equation VI.17. Note that u_s depends on the local stress field:

$$u_s = u_o + \frac{1}{2}\sigma_{ij}\varepsilon_{ij}v_o \ (v_o = \text{molar volume of the undeformed solid}).$$

When local stresses are too large, plasticity or microcracking can develop locally. The concept of a fluid film at grain contacts appears somewhat contradictory, because this film must behave as a fluid (allowing diffusion through solution) and as a solid (supporting the stress gradient which is the driving force). Such a contradiction would not exist if clay minerals are substituted for the fluid film at grain boundaries. The clay minerals can then provide short circuits for diffusion between highly stressed and stress-free areas. In such rocks, one observes dissolution at grain contacts and recrystallization at solid-fluid interfaces (fig. VI.8). This conceptual problem of a stress-bearing film can also be eliminated if the grain contacts are rough and the stress is maintained by asperities.

In order to avoid the difficulties of the previous theory, one can use a secondary analysis which is quite complementary. Instead of looking at the solid part of the porous rock, one can analyze the evolution of a cavity which is filled with a fluid at pressure p and is inside a solid medium under an isotropic external stress. A simple case is that of a cylindrical cavity of radius R (fig. VI.9). When the dissolution or

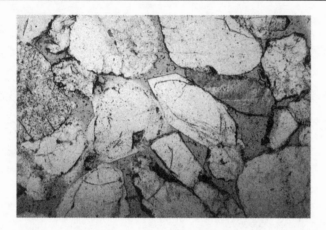

Fig. VI.8. Evidence of pressure solution in a Vosges sandstone (optical micrograph courtesy of D. Jeannette). Note areas of dissolution (at grain constants) and areas of recrystallization.

recrystallization of a layer δR occurs, the surface area of the fluid-solid interface varies.

In the more general case where the stress field is anisotropic (fig. VI.10), it is possible to calculate the total energy of the system (solid + fluid + loading system) and the results show that the energy reaches a minimum when the pore cross section is changed from a circular shape to an elliptical shape (Reuschlé et al., 1988). The energy minimum depends on the material properties (elastic moduli, surface energy), on the applied stresses (σ, ε, p), and on geometrical parameters (R). The results of simulations for different values of these parameters are shown in figure VI.11. Fluid pressure inhibits the process whereas σ activates it. At high σ values, energy decreases without a minimum ever being reached.

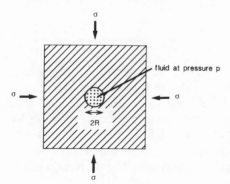

Fig. VI.9. Solid containing a cylindrical cavity and submitted to an isotropic stress σ.

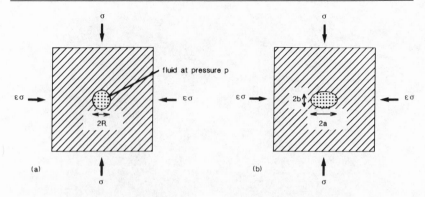

Fig. VI.10. Anisotropic stress field: (*a*) circular pore before pressure solution; (*b*) elliptical pore after pressure solution.

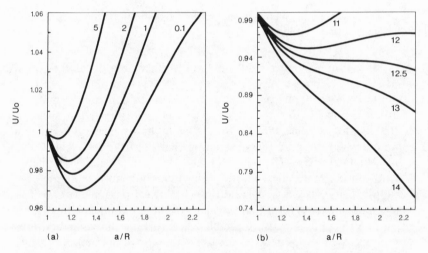

Fig. VI.11. Evolution of a cavity as a function of (*a*) fluid pressure p ($\sigma = 11$ MPa, $\varepsilon = 0$, $\gamma = 6$ Jm^{-2}, $E = 127$ GPa, $\nu = 0$, $R = 1$ mm); (*b*) vertical stress σ (values of parameters are as above with $p = 0.1$ MPa).

Kinetics and Rheological Laws

We have just seen why mass transfer through solution can take place. However, the kinetics of this process is also of fundamental importance, because it determines how long it will take for the total strain to take place. We know that solid-state diffusion is extremely slow under diagenetic conditions. It is also very slow for conditions of moderate metamorphism. Pressure solution is geologically important because diffusion through the fluid phase is relatively fast. The derivation of the

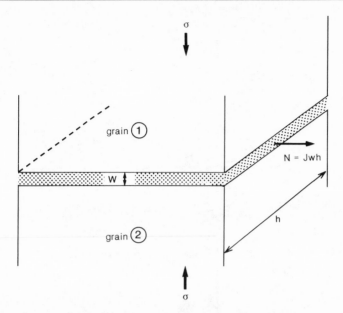

Fig. VI.12. Intergranular transport: the number of moles/s transported from a stressed face to a free face is $N = Jwh$ (w = thickness of intergranular film; h = grain length; J = flux).

rheological law $\dot{\varepsilon} = f(\sigma)$ is very similar to that presented for diffusion creep in chapter IV (equ. IV.24) with a few modifications: the molar concentration is substituted for vacancy concentration and diffusion takes place at grain boundaries. Then N, the number of moles transferred from a stressed face to a stress-free face is:

$$N = J \cdot w \cdot h$$

where w is the thickness of the intergranular zone (fig. VI.12). The concentration difference Δc between the two faces is

$$\Delta c \approx c_o \frac{\sigma v}{RT}$$

where v is the molar volume, R the universal gas constant, and c_o the equilibrium concentration of solid in solution (in moles m^{-3}). The above equation is an approximation to the following result for the chemical potential μ:

$$\mu = \mu_o + RT \log\left(\frac{c}{c_o}\right) \Leftrightarrow c = c_o \exp\left(\frac{\mu - \mu_o}{RT}\right)$$

with $\mu - \mu_o = \sigma v$ and $\sigma v \ll RT$. Substituting these relations into

IV.24, one finds:

$$\dot{\varepsilon} = \frac{\sigma \upsilon}{RT} \frac{D'w}{h^3} \qquad \text{VI.19}$$

D' is the diffusion coefficient through the intergranular fluid:

$$D' = D_g \bar{c}. \qquad \text{VI.20}$$

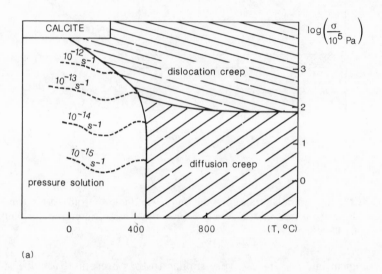

(a)

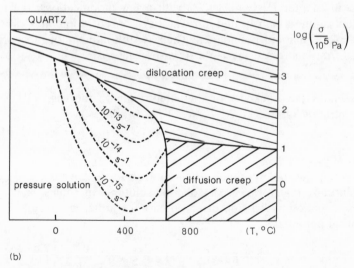

(b)

Fig. VI.13. Simplified deformation maps for: (*a*) calcite; (*b*) quartz. (From Rutter, E. H., 1976, The kinetics of rock deformation by pressure solution. *Phil. Trans. Royal Soc. London* A 283:203–19.

Here D_g is the diffusivity in the fluid along the interfaces and $\bar{c}$ the concentration of solid in solution (expressed as a volume fraction: $\bar{c} = c_o v$). Rutter (1976) has calculated that D_g is about 10^3 times less then D_v, the bulk diffusivity in the fluid. The exact value of D_g is however difficult to know. The numerical values used by Rutter to calculate the "deformation maps" for quartz and calcite shown in figure VI.13 are:

$$w = 2 \times 10^{-9} \text{ m} \quad h = 10^{-4} \text{ m} \quad v = 22 \text{ cm}^3 \text{ (quartz)}$$

$$v = 37 \text{ cm}^3 \text{ (calcite)}$$

$$D_g = 10^{-14} \exp\left\{ -\frac{E}{R}\left(\frac{1}{T} - \frac{1}{273}\right) \right\} \ m^2 s^{-1} \ (T \text{ in } °K).$$

Depending on the temperature conditions, $\bar{c}$ can vary from 10^{-5} to 10^{-3} in the domain [0–300 MPa; 0–500°C]. Figure VI.13 shows that, even at moderate conditions ($T = 200°C$, $\sigma = 10$ MPa), $\dot{\varepsilon}$ reaches a value $10^{-13} \ s^{-1}$ which is in the range of typical geological strain rates.

Such calculations, despite the fact that some of the material parameters are not well known, suggest strongly that pressure-solution plays a key role in crustal deformation. The progressive reduction of porosity observed at great depths is certainly controlled by this type of deformation mechanism.

PROBLEMS

VI.a. Time Constant for Fluid Diffusion

(1) Assuming a permeability value $k = 10$ nD for granite (see chap. V), calculate the value of the hydraulic diffusivity c. Give an estimate of the time constant τ for fluid diffusion to take place over a distance $l = 10$ km.

(2) Utilizing the results of chapter V, discuss the required conditions (in terms of crack density, extension, and aperture) in order to have $\tau \sim 1$ year.

VI.b. Poisson's Ratio in the Undrained Regime

(1) Show that in the undrained regime, $\nu_u = \dfrac{3\nu + \alpha B(1 - 2\nu)}{3 - \alpha B(1 - 2\nu)}$, where ν is the Poisson's ratio in the drained regime, B is the Skempton's coefficient, and α the effective pressure coefficient.

(2) Discuss the range of possible values for ν_u.

VI.c. Equivalent Medium and Poroelasticity

Use the results of chapter III to show how the bulk modulus K depends on a low concentration of spherical pores. Derive (1) K as a function of

(K_s, μ_s, ϕ), and (2) K_u as a function of (K_f, K_s, μ_s, ϕ). Show that both of these relations are compatible with equation VI.11.

VI.d. Pressure Solution in the Case of a Variable Surface Area

Assuming a spherical pore of radius R, show that relation VI.17 becomes: $\mu = f_s + p v_s - \gamma \left(\dfrac{v_s}{R} \right)$. γ is the surface energy.

REFERENCES

Atkinson, B. K. 1984. Subcritical crack growth in geological materials. *J. Geophys. Res.* 89:4077–4114.

Biot, M. A. 1941. General theory of three dimensional consolidation. *J. Appl. Phys.* 12:155–64.

Gratier, J. P. 1983. Estimation of volume changes by comparative chemical analysis in heterogeneously deformed rocks (folds with mass transfer). *J. Struct. Geol.* 5:329–39.

Jaeger, J. C., and N. G. Cook. 1979. *Fundamentals of Rock Mechanics.* London: Chapman and Hall.

Landau, L., and E. Lifchitz. 1967. *Statistical Physics.* Moscow: Editions Mir.

McTigue, D. F. 1986. Thermoelastic response of fluid saturated porous rock. *J. Geophys. Res.* 91:9533–42.

Michalske, T. A., and S. W. Freiman. 1981. A molecular interpretation of stress corrosion in silica. *Nature* 295:511–12.

Nur, A., and J. D. Byerle. 1971. An exact effective stress law for elastic deformation of rock with fluids. *J. Geophys. Res.* 76:6414–19.

Palciauskas, V. V., and P. A. Domenico. 1982. Characterization of drained and undrained response of thermally loaded repository rocks. *Water Res. Research* 18:281–90.

Reuschlé, T., L. Trotignon, and Y. Guéguen. 1988. Pore shape evolution by solution transfer: thermodynamics and mechanisms. *Geophys. J.* 95:535–48.

Rice, J. R., and M. P. Cleary. 1976. Some basic stress diffusion solutions for fluid saturated elastic porous media with compressible constituents. *Rev. Geophys. Space Phys.* 14:227–91.

Smith, D. L., and B. Evans. 1984. Diffusional crack healing in quartz. *J. Geophys. Res.* 89:4215–35.

VII. Acoustic Properties

BECAUSE of the important role of seismic prospecting and sonic logging, understanding the acoustic rock properties (velocity and attenuation) is essential for these and other geophysical applications. We have seen in chapter IV that the acoustic velocities V_p and V_s can be expressed simply in terms of the elastic constants and density of a medium. The elastic constants and density depend on lithology, porosity, fluid saturation, pressure, and temperature. Attenuation is an independent parameter which is potentially rich in information, but the many complex processes of attenuation are much more difficult to interpret. As a general rule, elastic anisotropy is not very pronounced, but there are certain cases where it must be taken into account.

VELOCITY OF ELASTIC WAVES

Velocity and Lithology

Lithology is an important factor which affects the velocities V_p and V_s. In *sedimentary rocks*, on the average, higher velocities are observed in carbonates than in sandstones and marl (fig. VII.1). But, unless we know their exact composition and/or their porosity, sandstone and calcite cannot be distinguished simply on the basis of V_p. Let us assume for the time being that the rocks are isotropic, homogeneous, and elastic. Then utilizing equations IV.8 and 9:

$$V_p = \sqrt{\frac{K + \frac{4}{3}\mu}{\rho}} \qquad V_s = \sqrt{\frac{\mu}{\rho}}.$$

The ratio V_p/V_s is interesting from the point of view of lithology because it only depends on the Poisson's ratio ν. With the aid of table IV.2, one can show

$$\nu = \frac{1}{2}\frac{(V_p/V_s)^2 - 2}{(V_p/V_s)^2 - 1}. \qquad \text{VII.1}$$

Figure VII.2 shows that the V_p/V_s ratio can be used to differentiate between sandstone and limestone. This result is partially a consequence

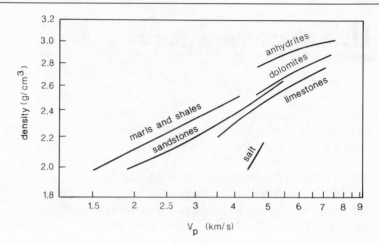

Fig. VII.1. Velocity V_p of the principal sedimentary rocks as a function of density. (After Gardner, G.H.F., L. W. Gardner, and A. R. Gregory, 1974. Formation velocity and density: the diagnostic basics for stratigraphic maps, *Geophysics* 39:370–80.)

of the very different values of the Poisson's coefficient for quartz and calcite:

$$\nu_{quartz} \approx 0.08 \qquad \nu_{calcite} \approx 0.31.$$

The V_p velocities of these two minerals, although quite different, cannot be used to identify the lithology due to the complicating effects of porosity and microstructure. Tables VII.1 and VII.2 illustrate this point for the principal sedimentary rocks. Comparison of these two tables shows that the rock velocity is primarily controlled by mineralogy, but only up to a certain point. Fluctuations in velocities arise due to the variable rock porosity, microstructure, and the presence of secondary minerals. This is particularly true for secondary minerals containing iron, when the velocity and density of the ensemble are increased. In general, there exists a correlation between velocity and density which is evident in figure VII.1 and table VII.2. The microscopic basis for this relation has been examined in chapter IV. This correlation implies that the three parameters are not independent and that there exist relations $K(\rho)$ for minerals of which equation IV.11 is a particular example. These relations $K(\rho)$ are parameterized by the average atomic mass $\overline{m}$, which gives distinct trajectories for anhydrite, sandstone, calcite, and salt in the $V_p - \rho$ plane as shown in figure VII.1. For sandstone $\overline{m} \approx 20$ while for salt, $\overline{m} \approx 29$.

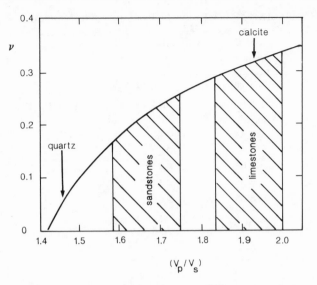

Fig. VII.2. V_p/V_s ratio as a function of Poisson's ratio ν.

Table VII.1. TYPICAL VALUES OF V_p FOR THE PRINCIPAL SEDIMENTARY ROCKS

Rock	Velocity V_p (km / s)
Limestone	4–7.0
Sandstone	2–5.5
Dolomite	5–7.5
Anhydrite	5–6.5
Shale	1.8–4.0
Clays	1.8–2.4
Salt	4.5

For the case of volcanic rocks, the fluctuations in porosity are very large and the velocity is principally controlled by the density, conforming to relation IV.11. One can present this correlation in the form velocity-density or velocity-acoustic impedance. The acoustic impedance $Z = \rho V_p$ appears explicitly in the expression for the reflection and transmission coefficients of waves. For example, for incident waves normal to the boundary between medium 1 and 2, the reflection coefficient is $R = \dfrac{Z_2 - Z_1}{Z_2 + Z_1}$. Figure VII.3 shows that, for volcanic and metamorphic rocks of identical average atomic mass, the correlation $Z - V_p$ is excellent.

Table VII.2. ACOUSTIC PARAMETERS OF ROCK FORMING MINERALS

Mineral	V_p (km / s)	V_s (km / s)	Density (g / cm³)	V_p / V_s
Salt	4.59	2.66	2.16	1.72
Anhydrite	5.63	3.11	3.00	1.81
Feldspar				
(K)	5.59	3.06	2.56	1.81
(Na)	5.94	3.27	2.62	1.82
(Ca)	7.05	3.73	2.73	1.89
Quartz	6.06	4.11	2.65	1.47
Calcite	6.65	3.45	2.71	1.93
Dolomite	7.37	3.99	2.87	1.85
Siderite	6.96	3.58	3.96	1.94
Pyrite	8.42	5.44	5.02	1.54

Porosity and Saturation Effects

The velocity of sedimentary rocks depends strongly on the rock porosity ϕ. A relation that is often utilized to account for porosity effects (for a rock saturated with fluid) is that proposed by Wyllie:

$$\frac{1}{V_p} = \frac{\phi}{V_f} + \frac{1 - \phi}{V_m} \qquad \text{VII.2}$$

where V_f = wave velocity in the fluid and V_m = wave velocity in the mineral aggregate ($\phi = 0$). This equation is often called the time average equation. As illustrated in figure VII.4, the equation assumes that transit time t for a wave through a rock is the sum of the time it travels through the fluid t_2 and through the solid t_1:

$$t = \frac{h}{V_p} = t_1 + t_2 = \frac{h_m}{V_m} + \frac{h_f}{V_f}.$$

With $h_f/h = \phi$ and $h_m/h = 1 - \phi$, one obtains Wyllie's equation. This is a simple model of the effective medium type which we have discussed in chapter III. Wyllie's equation is generally considered a reasonable approximation for cases where the porosity is intergranular (primary). It is not considered to be accurate when the porosity is secondary or for media with fracture porosity. Because the wavelength of elastic waves is larger than the grain or pore sizes, the waves do not travel only through the solid or the fluid, but through the aggregate system. Thus the time

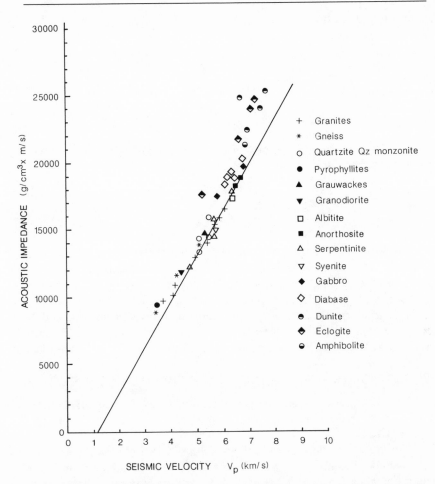

Fig. VII.3. Acoustic impedance $Z = \rho V_p$ as a function of V_p for the principal igneous and metamorphic rocks.

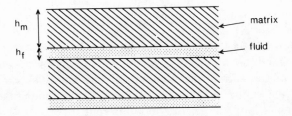

Fig. VII.4. A geometric model for interpreting the time average equation: the effective medium is considered as a medium of alternating layers.

Table VII.3. AVERAGE VALUES OF THE COEFFICIENTS IN THE
$V_s - \phi$ CORRELATION

	$A\ (10^{-6}\ sec\ m^{-1})$	$B\ (10^{-6}\ sec\ m^{-1})$
Sandstone	230	1000
Limestone	316	400

average model does not represent the appropriate microstructure of sedimentary rocks.

Wyllie's relation can also be written as

$$\frac{1}{V} = A + B\phi. \qquad \text{VII.3}$$

In this form the equation is very general, but the interpretation of the constant B is not as direct as in the time-averaged equation. In this form A is seen to be the inverse of the velocity of the mineral aggregate at zero porosity. B depends on many factors such as pore geometries, compaction, pressure, etc. Average values for the coefficients A and B for shear waves are presented in table VII.3 for sandstone and limestone (fig. VII.5). $1/V_s$ is proportional to ϕ for sandstone. For the case of limestone, it is found that $\left(\dfrac{1}{V_s} - \dfrac{1}{V_p} \right)$ is proportional to ϕ. Equation VII.3 can be derived from the basic relations presented in chapters IV and VI (in particular equs. IV.12 and VI.11). These relations are summarized in table VII.4.

K_s and μ_s are the rock moduli at zero porosity (mineral moduli), K_f is the bulk modulus of the fluid, ρ_s and ρ_f are the mineral and pore fluid densities. The constants β and β' are approximately unity. A is the pore shape factor ($A \leq 1$). The results in table VII.4 show that when ϕ increases for a saturated medium, K_u and μ decrease and so does ρ'. But the decrease of K_u and μ is more rapid than that of ρ'. Thus to the leading order of approximation, the velocities have the form as in equation VII.3, and B is expressed in terms of the constants appearing in table VII.4.

In real situations a porous medium is not always fully saturated. It can also contain several different fluids. The velocities vary with the nature of the saturating fluid and degree of saturation. Because the density ρ_f and modulus K_f are very different for gases ($K_f \sim 10^{-6}$ to 2.10^{-4} M bars) and liquids ($K_f \sim 10^{-2}$ M bars), the velocity V_p and the acoustic impedance Z decrease significantly when only a small fraction of gas is present. If the medium does not contain fluids, the modulus K_u

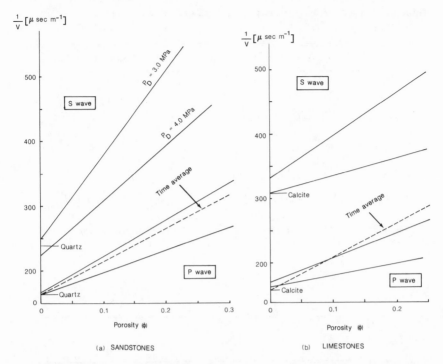

Fig. VII.5. $V^{-1} - \phi$ relations for sandstones and calcites. The curves correspond to differential pressures of 40 and 3 MPa. (After Domenico, P. A., 1984, Rock lithology and porosity determination from shear and compressional wave velocity. *Geophysics* 49:1188–95.)

is replaced by K (the drained or frame modulus). The variations in V_p and V_s with degree of saturation S are relatively complex because many effects are superimposed.

If one compares the velocities V_p and V_s in a dry medium ($S = 0$) and one saturated with water ($S = 1$), one observes that V_p (dry) $< V_p$ (saturated) and V_s (saturated) $< V_s$ (dry). This observation can be interpreted through the results in table VII.4. Saturation does not change the shear modulus μ, but increases the bulk rock density, thus V_s (saturated) $< V_s$ (dry). But saturation also increases the frame modulus ($K < K_u$), usually by a greater relative amount than the increase in density, thus V_p (dry) $< V_p$ (saturated). Figure VII.6 illustrates these effects and presents evidence that the ratio (V_p/V_s) can indicate incomplete rock saturation, or the presence of gas. The shape of the curves indicates that the velocities V_p and V_s vary with saturation in a complicated manner. Unfortunately, table VII.4 only allows us to discuss the

Table VII.4. BASIC RELATIONS BETWEEN ACOUSTIC, ELASTIC, AND
MICROSTRUCTURAL PARAMETERS OF A POROUS MEDIUM

	Porous Medium without Fluid	Porous Medium with Fluid
Bulk Modulus	$K = K_s\left(1 - \beta\dfrac{\phi}{A}\right)$	$K_u = K + \dfrac{K_f\left(1 - \dfrac{K}{K_s}\right)^2}{\phi + \left(1 - \dfrac{K}{K_s} - \phi\right)\dfrac{K_f}{K_s}}$
Shear Modulus	$\mu = \mu_s\left(1 - \beta'\dfrac{\phi}{A}\right)$	$\mu_u = \mu$
Density	$\rho = \rho_s(1 - \phi)$	$\rho' = \rho_s(1 - \phi) + \rho_f\phi$
P = Wave	$V_p = \left(\dfrac{K + (4/3)\mu}{\rho}\right)^{1/2}$	$V_p = \left(\dfrac{K_u + (4/3)\mu}{\rho'}\right)^{1/2}$
S-Wave	$V_s = \left(\dfrac{\mu}{\rho}\right)^{1/2}$	$V_s = \left(\dfrac{\mu}{\rho'}\right)^{1/2}$

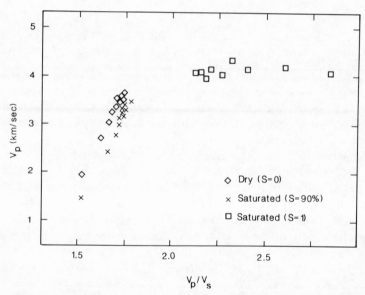

Fig. VII.6. V_p as a function of V_p/V_s for various degrees of saturation. Massilon
sandstone. (After Winkler, K. W., and A. Nur, 1982, Seismic attenuation: effects
of pore fluids and frictional sliding, *Geophysics* 1:1–15.)

two limiting saturations $S = 0$ and 1. If the porosity is large, the initial saturation $(S < 1)$ affects the density more than the modulus, and V_p and V_s decrease. When the saturation approaches one, the effect on the modulus becomes dominant and V_p increases. Thus there is a minimum in the curve $V_p(S)$ due to the different ways saturation affects modulus and density. The magnitude of these effects depends on the porosity structure and can be particularly pronounced in rocks with fracture porosity. Fracture porosity is usually very small and the effect of saturation on density is negligible. But if the pore shape factor A is very small, then the change in the moduli K and μ can be quite large due to their $1/A$ dependence.

Variation with Pressure and Temperature

The velocities V_p and V_s decrease when the temperature increases, but the magnitude of decrease is quite small as can be seen from table VII.5. On the average, the decrease in velocities does not exceed 5% for a temperature change of 100°C.

The effect of pressure is more important. As for the permeability, we must distinguish between the pore pressure p and the confining pressure P. Figure VII.7 shows how V_p and V_s vary with p and P: the rapid increase in velocities between $P = 0$ and $P = 100$ MPa (at $p = 0$) is attributed to the closing of cracks, as suggested by the relation IV.14. A crack with shape factor $A = 10^{-3}$ is closed when the confining pressure is $P \approx AE$. If the Young's modulus is $E = 50$ GPa, this will occur at a confining pressure $P \approx 50$ MPa. The pore pressure has the opposite effect to the confining pressure. One can define the "effective" pressure in a sense analogous to the definition for permeability (equs. V.7 and 8):

$$V(P_{\text{eff}}, p = 0) = V(P, p)$$

which results in the approximate relation:

$$P_{\text{eff}} = P - np.$$

Table VII.5. V_p (KM S^{-1}) AS A FUNCTION OF T, AT $P = 50$ MPA

T(°C)	25	100	200
Sandstone (dry, $\phi = 0.05$)	4.58	4.38	4.14
Calcite (Solenhofen)	6.03	5.93	5.78
Basalt	5.57	5.59	5.50
Gabbro (San Marcos)	6.79	6.78	6.73
Granite (Texas)	6.29	6.23	5.86

Source: After Christensen, N. I., 1982, Seismic velocities. In R. S. Carmichael, ed., *Handbook of Physical Properties of Rocks*. Boca Raton, Fla.: CRC Press.

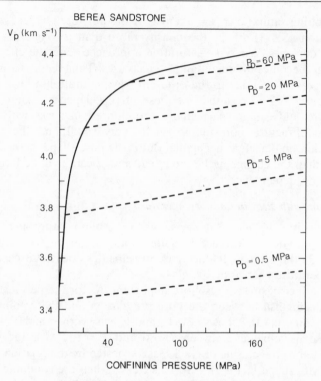

Fig. VII.7. Variation of V_p. with confining pressure (P) and pore pressure (p). $P_D = P - p$ is the differential pressure. Berea sandstone. (After Christensen, N. I., and H. F. Wang, 1985, The influence of pore pressure and confining pressure on dynamic elastic properties of Berea sandstone, *Geophysics* 50:207–13.)

One observes the same type of behavior for crystalline rocks (fig. VII.8): a rapid increase in the "extrinsic" regime associated with the closure of cracks, followed by a slow increase in the "intrinsic" regime.

ATTENUATION OF ACOUSTIC WAVES

The essential hypothesis of elasticity is that the deformations are reversible. Hook's law implies that for a given stress σ, there exists a unique deformation ε such that $\varepsilon = \sigma/M$ (equ. IV.5), and that if the stress is removed, the total deformation is zero and the stored elastic energy is completely recovered. Generally, this hypothesis is accurate for small deformations. As far as acoustic waves are concerned, the amplitudes of deformation are very small and irreversible effects are ignored in the leading approximation. But there is always a small

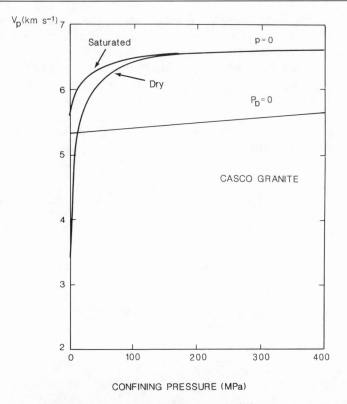

Fig. VII.8. Variation of V_p with confining pressure (P) and pore pressure (p). Casco granite. (After Nur, A., 1980, Seismic velocities in low porosity rocks: source mechanism and earthquake prediction, CNRS, Paris.)

amount of energy dissipated. This dissipation results in the reduction of the wave amplitude with distance according to an exponential law:

$$A = A_o e^{-\alpha x}. \qquad \text{VII.4}$$

The energy of a wave is proportional to the amplitude squared ($\sim A^2$) and thus decreases with distance x as $e^{-2\alpha x}$.

Viscoelastic Models

The dissipation of energy occurs when stresses and displacements are out of phase by an angle $\varphi \neq 0$. Linear viscoelasticity is a generalization of elasticity which permits one to take account of this phase difference through the notion of a complex elastic modulus $\overline{M}(\omega)$. $\overline{M}(\omega)$ is a function of the angular frequency ω:

$$\overline{M} = M_1 + iM_2. \qquad \text{VII.5}$$

Here we utilize the complex notation for a time-dependent strain $\varepsilon = \varepsilon_o e^{i\omega t}$ (as for the alternating current in electricity). The real deformation is described by the real part of the preceding expression. The notion of complex moduli is particularly useful for sinusoidal waves. Hooke's law (equs. IV.6) remains applicable, with the complex moduli $\overline{K} = K_1 + iK_2$ and $\overline{\mu} = \mu_1 + i\mu_2$ replacing the real moduli K and μ of elasticity.

The constitutive relations of viscoelasticity (for sinusoidal waves) can be written in the form:

$$\varepsilon_{kk} = \frac{\sigma_{kk}}{3\overline{K}}$$

and for $i \neq j$:   VII.6

$$\varepsilon_{ij} = \frac{\sigma_{ij}}{3\overline{\mu}}.$$

The moduli $\overline{K}$ and $\overline{\mu}$ are functions of ω. If $\varepsilon = \varepsilon_o e^{i\omega t}$ and $\sigma = \overline{M}\varepsilon$, the energy dissipated in one cycle is

$$\Delta W = \int_{\omega t = 0}^{2\pi} \sigma \, d\varepsilon = \pi M_2 \varepsilon_o^2.$$

The maximum stored elastic energy is:

$$W = \int_{\omega t = 0}^{\pi/2} \sigma \, d\varepsilon = \frac{M_1}{2} \varepsilon_o^2.$$

(For the last calculation, it is necessary to take only the real parts of σ and ε which represent the elastic behavior.) Thus the fractional energy dissipated in one cycle is:

$$\frac{\Delta W}{W} = 2\pi \frac{M_2}{M_1} \equiv 2\pi \tan \varphi.$$ VII.7

It has become standard to describe the average viscoelastic response of solids by the simple analog models shown schematically in figure VII.9. The three models of viscoelastic behavior are composed of only two types of elements: elastic spring with force constant M and a dashpot with viscosity η_b (representing viscous behavior). These three classic models are the ones most often utilized. Due to the different interconnections, the frequency response of $\overline{M}(\omega)$ is different for the three cases. The expressions for $\overline{M}(\omega)$ for these three models are presented in table VII.6. These expressions can be obtained by noting that elements in series have equal stresses (deformations are additive), while elements in parallel have equal deformations (stresses are additive). The response of the elastic elements is described by the relation

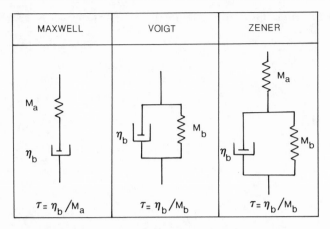

Fig. VII.9. The Visco-elastic models of: (a) Maxwell; (b) Voigt; and (c) Zener.

Table VII.6. Modulus $\overline{M}$ for the three Examples Shown in Figure VII.9

	Maxwell	Voigt	Zener
$\overline{M} = M_1 + iM_2$	$\dfrac{i\omega\tau}{1 + i\omega\tau}M_a$	$(1 + i\omega\tau)M_a$	$\dfrac{1 + i\omega\tau}{\left(\dfrac{1}{M_a} + \dfrac{1}{M_b}\right) + \dfrac{i\omega\tau}{M_a}}$

$\sigma = M\varepsilon$, those of the viscous elements by $\sigma = \eta\dot{\varepsilon}$. For example, Maxwell's model implies:

$$\begin{array}{ll} \varepsilon = \varepsilon_a + \varepsilon_b & \sigma_a = M_a\varepsilon_a \\ \sigma = \sigma_a = \sigma_b & \sigma_b = \eta_b\dot{\varepsilon}_b \end{array} \Rightarrow \dot{\varepsilon} = \frac{\dot{\sigma}}{M_a} + \frac{\sigma}{\eta_b}.$$

With $\varepsilon = \varepsilon_o e^{i\omega t}$ and $\sigma = \overline{M}\varepsilon$, one obtains

$$\overline{M} = \frac{i\omega\tau}{1 + i\omega\tau}M_a \quad \text{where } \tau = \frac{\eta_b}{M_a}.$$

Maxwell's model yields an elastic response at high frequencies ($\omega\tau \gg 1$) and a viscous fluid response at low frequencies ($\omega\tau \ll 1$). τ is the characteristic time constant for the system.

Q Factor

The quality factor Q is a measure of the elastic energy lost during one period of oscillation. It is defined as

$$Q^{-1} = \frac{1}{2\pi}\frac{\Delta W}{W}. \qquad \text{VII.8}$$

The preceding definition in conjunction with VII.7 implies that $Q^{-1} = \tan \varphi = \dfrac{M_2}{M_1}$. For the case of elastic wave attenuation, one usually finds $M_2 \ll M_1$, and $Q^{-1} \ll 1$. Thus for a plane wave $u = u_o e^{i\omega(t-x/V)}$, with $V = \left(\dfrac{\overline{M}}{\rho}\right)^{1/2} \cong \left(\dfrac{M_1}{\rho}\right)^{1/2}\left(1 + \dfrac{i}{2}\dfrac{M_2}{M_1}\right)$, by defining $V_o = \left(\dfrac{M_1}{\rho}\right)^{1/2}$ and $Q^{-1} = \dfrac{M_2}{M_1}$ one obtains

$$u = u_o e^{-\alpha x} e^{i\omega(t-x/V_o)}$$

where

$$\alpha = \frac{\omega Q^{-1}}{2V_o}. \qquad\qquad \text{VII.9}$$

The frequency dependence of the Q factor varies with the model chosen. For the case of Zener's model, one obtains peak energy loss (absorption) at $\omega\tau = 1$ (fig. VII.10a). In effect, with the approximation $M_a \ll M_b$, the result in table VII.6 can be simplified to:

$$Q^{-1} \cong \frac{\omega\tau}{1 + \omega^2\tau^2}\frac{M_a}{M_b}.$$

At high frequencies ($\omega\tau \gg 1$) or low frequencies ($\omega\tau \ll 1$), absorption is negligible. Peak absorption is accompanied by a variation of velocity because the modulus M is equal to $\left(\dfrac{1}{M_a} + \dfrac{1}{M_b}\right)^{-1}$ at low frequencies and M_b at high frequencies: V increases with ω.

If one superposes many Zener models which correspond to a distribution of relaxation times τ over an interval $[\tau_1, \tau_2]$, one obtains a succession of absorption peaks and Q^{-1} is approximately constant over the interval $[\tau_1, \tau_2]$ (fig. VII.10b). The velocity increases over this interval (fig. VII.10c). Calculation shows that the velocity dispersion follows the law:

$$\frac{V(\omega)}{V(\omega_o)} \cong 1 + \frac{1}{\pi Q}\ln\frac{\omega}{\omega_o}.$$

Physical Processes

Experimental data show that for *sedimentary rocks*, attenuation depends strongly on both the fluid saturation and differential pressure (fig. VII.11). Experimental data are usually obtained by the method of low frequency resonance (Bourbié, Coussy, and Zinzer, 1986) over the range

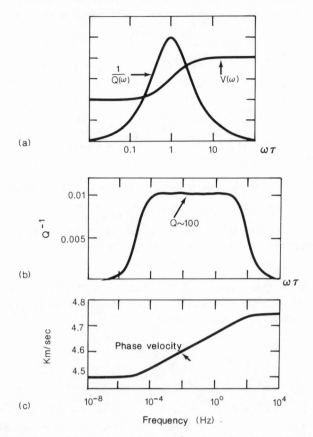

Fig. VII.10. Zener model: variation of Q^{-1} and V with angular frequency ω.

$5 \cdot 10^2$ Hz–10^4 Hz. Attenuation is minimal in a dry medium for both P and S waves. At total saturation, attenuation of S waves is maximum, but for P waves it is still rather weak. The attenuation of P waves is most strong when the rock is partially saturated. These experimental observations can be explained in terms of local fluid-flow processes. They are more pronounced in a compressional regime when the saturation is incomplete, while they are always possible during shear deformation, thus the results of figure VII.11.

Temperature effects are much more important for *crystalline rocks*. The physical processes causing attenuation are also very different. Thermoelastic relaxation, grain boundary sliding, and the mass movement of dislocations are the principal processes of dissipation. Figure VII.12 shows the variation of attenuation in a tholeitic basalt: the two peaks observed are attributed to grain boundary sliding. The bimodal

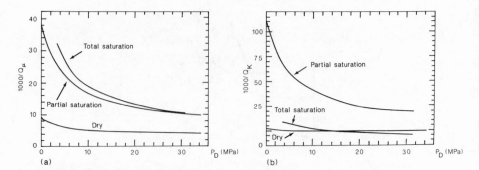

Fig. VII.11. Massilon sandstone ($\phi = 0.23$). Attenuation in the shear (a) and compressional (b) regimes. $Q_\mu = \dfrac{\mu_2}{\mu_1}$ and $Q_k = \dfrac{K_2}{K_1}$. P_D is the differential pressure ($P - p$). (After Winkler, K. W., and A. Nur, 1982, Seismic attenuation: effects of pore fluids and frictional sliding, *Geophysics* 47:1–15.)

distribution of grain sizes and presence of a jointed vitreous phase constitute good arguments in favor of this mechanism. The absorption peaks correspond to a time constant

$$\tau \cong \frac{1}{A} \frac{\eta}{E}$$

where A is the joint shape factor (A = aperture width/length), η the viscosity of the vitreous phase, and E the Young's modulus of the solids composing the rock. Viscosity η varies strongly with temperature such that a sweep of a temperature range is in a certain way equivalent to a sweep of a frequency range.

The behavior of a single crystal of forsterite (fig. VII.13) is very different. By comparing deformed and undeformed crystals, it is possible to observe clearly the effect dislocations have on attenuation. There is no longer a peak but a continuous increase of Q^{-1} at high temperatures. The attenuation follows the law:

$$Q^{-1} = A \left[\omega \exp\left(\frac{E}{RT} \right) \right]^{-n}$$

where $n = 0.20$ and $E = 440$ kJ mole^{-1}. The activation energy observed here is close to that observed in creep for the same mineral (see chap. IV).

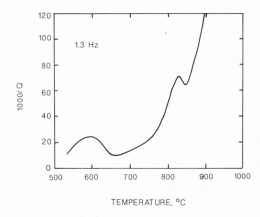

Fig. VII.12. Tholeitic basalt. Measured attenuation in a vacuum at 1.3 Hz with a pendulum. (After Weiner, A. T., M. H. Manghani, and R. Raj, 1987, Internal friction in tholeitic basalts, *J. Geophys. Res.* 92:635–43.)

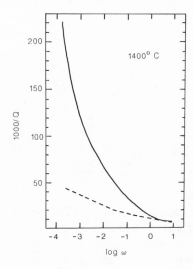

Fig. VII.13. Single crystal of Forsterite Mg_2SiO_4. Attenuation was measured at 1400°C in a vacuum with a pendulum. Solid line: sample deformed at 1600°C. Dashed line: undeformed sample. (After Guéguen, Y., and M. Darot, 1989, Q^{-1} for forsterite single crystals. *Phys. Earth and Plan. Int.* 55:254–58.)

VELOCITY ANISOTROPY

Minerals are always anisotropic from the point of view of elasticity. Rocks, on the other hand, are in general isotropic or weakly anisotropic. This apparent contradiction can be explained quite simply. For an elastic wave, the relevant length scale is its wavelength. Effective elastic properties involve an average over N grains which are oriented randomly (N being very large). Thus if an anisotropy appears, it is due to the fact that the distribution of grains (or porosity) is not random but has a preferred direction (the orientation of cracks, for example). Velocity anisotropy is generally small, less than 10% for most rocks.

Models of Anisotropic Media

Elasticity of an isotropic media is described by two constants (K and μ, or E and ν). We have seen in chapter IV that in the general case of anisotropic media twenty-one elastic constants C_{ijkl} (or C_{ij} using the simplified notation) are required. Between these two extremes—two constants or twenty-one—the laws of symmetry restrict the number of possible situations to seven (Landau and Lifchitz, 1967), two of which are particularly interesting and occur quite often.

The first corresponds to hexagonal symmetry for which there are five non-zero elastic constants. For this case there exists a two-dimensional isotropy in the plane where all axes are equivalent (rotational symmetry around the z-axis), but the direction perpendicular to this plane is not equivalent to any of the axes in the plane. Often this type of medium is called "transversely isotropic." The non-zero coefficients are (relation IV.18)

$$\begin{bmatrix} C_{11} & C_{12} & C_{13} & 0 & 0 & 0 \\ C_{12} & C_{11} & C_{13} & 0 & 0 & 0 \\ C_{13} & C_{13} & C_{33} & 0 & 0 & 0 \\ 0 & 0 & 0 & C_{44} & 0 & 0 \\ 0 & 0 & 0 & 0 & C_{44} & 0 \\ 0 & 0 & 0 & 0 & 0 & \dfrac{C_{11} - C_{12}}{2} \end{bmatrix}.$$

The P wave velocity is the same for all directions in the xy-plane but is different in the z-direction. S wave velocities in the symmetry plane depend on their polarization (vibration in the plane or perpendicular to it): birefringence exists for all directions in the xy-plane. Figure VII.14 summarizes these results.

A second type of anisotropy, one much more complex, corresponds to orthorhombic symmetry. In this case all three coordinate axes are not equivalent: thus $C_{13} \neq C_{23}$, $C_{44} \neq C_{55} \neq C_{66}$. Because the two axes OX and OY (fig. VII.14) are no longer equivalent, there exist two distinct velocities V_s for each direction.

Anisotropy Due to Fractures

A change in the degree of anisotropy with confining pressure is a good indication that fractures are present. In the example shown in figure VII.15, the anisotropy of amphibole and peridotite remains at 15% and 5%, respectively, at a pressure of 600 MPa, while the anisotropy of

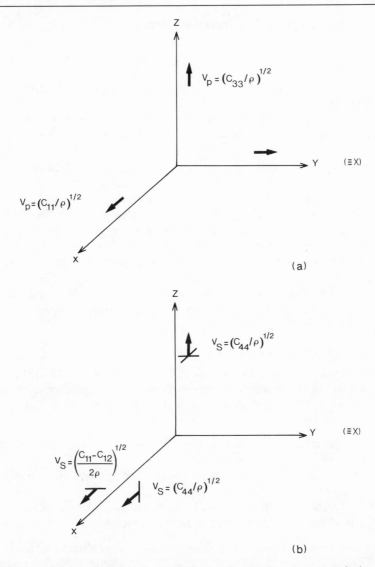

Fig. VII.14. Transversely isotropic medium: the Ox and Oy axes are equivalent. Oz is the axis of rotational symmetry. (a) Velocity of P waves; (b) Velocity of S waves.

granite and granulite becomes almost negligible at this pressure. If one assumes that for the last two rocks, the crack closure pressure is 200 MPa and Young's modulus is 50 GPa, relation IV.15 gives us the following estimate for the shape factor of the cracks: $A \sim 4 \cdot 10^{-3}$. Furthermore, we have seen that the bulk modulus K and velocity V_p

vary with the ratio ϕ/A (equ. IV.12). Thus:

$$\frac{\Delta V_p}{V_p} \cong \frac{1}{2}\frac{\Delta K}{K} \cong \beta\frac{\phi}{2A}$$

where β is a constant with a value close to one. A decrease in anisotropy of 5% between 0 and 200 MPa corresponds to a fracture porosity of $\phi = 0.4 \ 10^{-3}$. The total fracture porosity must be higher, because the preceding value represents fractures that were oriented non-randomly.

Applied stress fields will thus cause a departure from isotropy when the medium has an isotropic distribution of fractures. Depending on the symmetry of the applied stress field, the resulting anisotropy can either be transversely isotropic or orthorhombic. One finds the first case for uniaxial compression and the second case for more general applied stress fields. In most cases, fracture anisotropy is not possible for small differential pressures ($P < 200$ MPa) because these pressures correspond to a very small total deformation ($\ll 10^{-2}$).

Fracturing may be one of the causes of a measurable anisotropy in the upper crust and perhaps also in the lower crust, when the fluid pressure is high.

Anisotropy Due to Plastic Deformation

The residual anisotropy observed at high pressures (fig. VII.15) cannot be explained by fractures. It results from the preferred orientation of the rock-forming minerals. The existence of anisotropy similar to this example requires several favorable conditions:

(1) the principal rock-forming mineral possesses a significant intrinsic anisotropy;
(2) that there exist a process to orient this mineral in a preferred direction (non-randomly);
(3) if several minerals of the rock satisfy the two preceding conditions, their effects should not neutralize each other.

These conditions are satisfied by peridotite (fig. VII.15), which is the principle mineral of the upper mantle. Peridotite contains approximately 60% olivine and 35% pyroxene. Olivine, which has orthorhombic symmetry, is elastically anisotropic and exhibits marked plasticity. Under mantle conditions, pyroxenes are not very ductile and thus plastic deformation occurs mainly due to olivine. At high temperature, creep in olivine occurs preferentially along the slip system (010) [100] the preferred orientation: the crystals tend to align their [100] axes along the direction of shear. This slip direction [100] is also the direction of maximum V_p. Figure VII.16 shows how V_p^{max} and V_p^{min} vary for

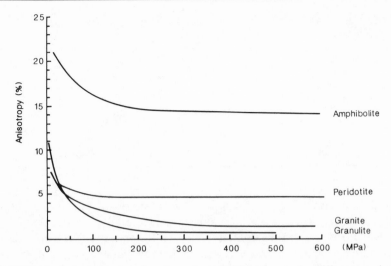

Fig. VII.15. Anisotropy as a function of confining pressure for different rocks. (After Kern, H., 1978, The effect of high temperature and high confining pressure on compressional wave velocities in quartz bearing and quartz free igneous and metamorphic rocks, *Tectonophysics* 44:185.)

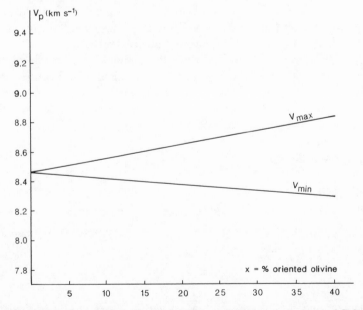

Fig. VII.16. Peridotite containing 60% olivine and 40% orthopyroxene. Calculation of V_p^{max} and V_p^{min} as a function of the fraction x of aligned olivine. The velocities are computed from the single crystal elastic constants at mantle conditions at 200 km ($T = 1200°C$, $P = 67$ kb).

Table VII.7. VELOCITY ANISOTROPY FOR THE PRINCIPAL MINERALS OF THE EARTH'S CRUST

Mineral	% anisotropy in V_p
Calcite	0.33
Quartz	0.26
Plagioclase	0.31
Alkali feldspar	0.46
Hornblende	0.27

peridotite containing 60% olivine and 40% orthopyroxene, as a function of the olivine fraction oriented along the axis [100].

For crustal rocks the situation is quite different. The first of the three conditions is almost always satisfied: most of the minerals are anisotropic as can be seen in table VII.7. But it is more difficult to satisfy conditions (2) and (3). Plastic deformation probably does not take place except at high temperature (i.e., in the deep crust) and/or in localized shear zones, as discussed in chapter IV. Moreover, because of the complex composition of rocks, it is not obvious that a single mineral will impose its anisotropy on the whole rock. The case of amphibolite presented in figure VII.15 is not typical of an average crustal rock.

Anisotropy due to plastic flow is the result of a large finite strain (greater than 50%), so that it cannot be developed on a short time scale. Plastic flow is a slow process which can eventually produce anisotropy in the deep crust and in the mantle. These characteristics are very different from those of crack-induced anisotropy.

PROBLEMS

VII.a.

Utilizing the results presented in table VII.4, calculate the coefficient B in the equation $\frac{1}{V} = A + B\phi$. Compare this result with Wyllie's equation.

VII.b.

Calculate the moduli M_1 and M_2 for the Zener model (utilize table VII.6). Find an expression for Q^{-1} in the approximation that $M_a \ll M_b$. Show that $(Q^{-1})_{max} \approx \left(\frac{\Delta V}{V}\right) \approx \frac{M_a}{2M_b}$.

VII.c. Elastic Wave Propagation

(1) Derive the equation for elastic wave propagation for the case of orthorhombic symmetry. Assume that the displacement vector has the form $\vec{u} = \vec{a}\exp\{i(\omega t - \vec{q}.\vec{r})\}$, where $\vec{a}$ is a constant vector which defines the wave polarization: $\omega = 2\pi/T$ where T is the period of oscillation, $|\vec{q}| = 2\pi/\lambda$ where λ is the wavelength, and $\omega/q = V$ is the velocity of propagation. Show that the propagation equation reduces to an eigenvalue equation defined by the matrix M and determine the components of M. The eigenvalues are the allowable values of ω.

(2) Consider the case of longitudinal polarization. Show that the velocity of P-waves is different for the three reference directions. Determine numerical values for olivine using $C_{11} = 323$ GPa, $C_{22} = 197.6$ GPa, $C_{33} = 236.1$ GPa (standard conditions). Compare with figure VII.16.

REFERENCES

Bourbie, T., O. Coussey, and B. Zinzer 1986. *Acoustics of porous media*. Paris: Edit. Technip.

Christensen, N. I. 1982. Seismic velocities. In R. S. Carmichael, ed., *Handbook of Physical Properties of Rocks*. Boca Raton, Florida: CRC Press.

Landau, L., and E. Lifchitz 1967. *Theory of elasticity*. Moscow: Editions Mir.

VIII. Electrical Conductivity

THE INTERPRETATION of electric field data measured at the Earth's surface requires knowledge of the electrical properties of rocks comprising the Earth's interior. Conductivity data are important for the interpretation of measurements of natural electromagnetic (EM) induction in the sea floor from ionospheric currents, electric fields from ocean currents and tides, and of artificial sources of electromagnetic induction. The electrical properties of rock are now being used routinely in applications such as well logging, mining, prospecting for mineral deposits, and other purposes in geology and geophysics. For example, electrical resistivity measurements were first used in 1963 to outline the Wairekei geothermal field in New Zealand. Mineral exploration surveys have been a prime application of electromagnetic methods, due to the fact that most base metal sulfide ores are very conductive as compared to the host rocks. In recent years there have been new developments and increased usage of electrical methods, all of which require a knowledge of how sedimentary and igneous rocks respond to the variety of electromagnetic field sources.

The conductivity of a pure metal at low temperatures may be on the order of 10^{12} S/m, while the conductivity of a good insulator may be as low as 10^{-20} S/m. This observed range of 10^{32} in conductivity may be the widest range of any common physical property of solids. Rocks and rock-forming minerals are also quite varied in their electrical properties with their conductivities ranging from 10^4 to 10^{-14} S/m. Such a large contrast implies that EM methods can be powerful tools in the detection of areas of anomalous conductivity.

LAWS AND PHYSICAL PROCESSES

In this chapter we shall focus on the electrical conduction processes in minerals and fluids at low frequencies ($\omega < 10^3$ Hz). The conductivity at higher frequences will be discussed in chapter IX, with polarization phenomena.

Ohm's Law

The electrical resistivity ρ and its inverse the electrical conductivity σ are quantities which characterize electrical charge transport. When a static electric field $\vec{E}$ is applied, an electric current density $\vec{J}$ is established due to the displacement of various charged particles, such as electrons and/or ions. The linear relation

$$\vec{J} = \sigma \vec{E} = \frac{1}{\rho} \vec{E} \qquad\qquad \text{VIII.1}$$

defines the electrical conductivity/resistivity of an isotropic material. $\vec{E}$ is expressed in volts/m and $\vec{J}$ in amperes/m². Equation VIII.1 is an expression of Ohm's law. When a material is anisotropic, VIII.1 is replaced by:

$$J_i = \sigma_{ij} E_j.$$

The conductivity σ and resistivity ρ are intrinsic material properties, independent of the sample geometry. Resistivity is generally expressed in units of ohm-meters while the conductivity is expressed in the reciprocal unit siemens/meter (or mhos/meter). The electrical resistance R (ohms) of a sample depends on its geometry. For a homogeneous sample of uniform cross-section A and length L, the resistance is

$$R = \rho \left(\frac{L}{A} \right)$$

when the field is applied perpendicular to the cross-sectional area A.

In a static electric field $\vec{E}$, the force on a charge q is $q\vec{E}$. At steady state, this results in charge displacement with a uniform velocity, $\vec{v}$.

$$\vec{v} = \mu \vec{E}. \qquad\qquad \text{VIII.2}$$

μ is the mobility of the charge and depends on particle type and its interaction with the medium. The electric current density $\vec{J}$ is the amount of charge that crosses a unit surface area per second. If n is the number of charge carriers per unit volume,

$$\vec{J} = nq\vec{v}. \qquad\qquad \text{VIII.3}$$

Combining equations VIII.2 and 3 yields an expression for the conductivity.

$$\sigma = nq\mu. \qquad\qquad \text{VIII.4}$$

The conductivity is proportional to the density of charge carriers, their charge, and their mobility. Equations VIII.4 is a typical transport equation, analogous for example to equation IV.25.

On the basis of their electrical conductivity, solids are generally divided into three groups: conductors ($10^5 < \sigma < 10^8$ S/m), semiconductors ($10^{-7} < \sigma < 10^5$ S/m), and insulators ($\sigma < 10^{-7}$ S/m). These

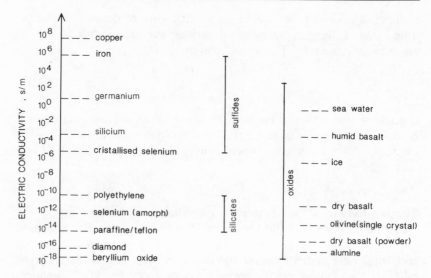

Fig. VIII.1. DC conductivity of various materials at room temperature. Note the importance of water and the nature of the medium (e.g., powdered materials). (After Olhoeft, G. R., 1981, Electrical Properties of Rocks, in *Physical Properties of Rocks and Minerals*, ed. Touloukian, Y. S., W. R. Judd, and R. F. Roy, vol. II-2 McGraw-Hill [Cindas Data Series on Material Properties].)

extreme variations of conductivity reflect the variations in the nature of the charge carrier, their number density, and their mobility (figs. VIII.1 and 2). Note that the conductivity of ocean water is approximately 10 orders of magnitude larger than that of silicate minerals, the major constituent of crustal rocks. It is thus not surprising that even modest amounts of water in porous silicate rocks can dramatically increase their conductivity.

Electronic Conduction

Although every solid contains electrons, the electrical conductivity varies by many orders of magnitude from one solid to another. In crystals, electron energy states are arranged in discrete bands which are separated by disallowed energy regions (fig. VIII.3). When a band is only partially filled, the electrons can access a continuum of states of higher energy. For this reason they can respond to electric fields and acquire a drift velocity through the crystal. This is the situation for metals and is the reason why they are good conductors. In insulators the number of electrons is such that one or more bands are completely filled and a large energy gap E_g exists to the next empty band (fig. VIII.3). Due to the large energy gap, essentially no electrons can reach the next-higher

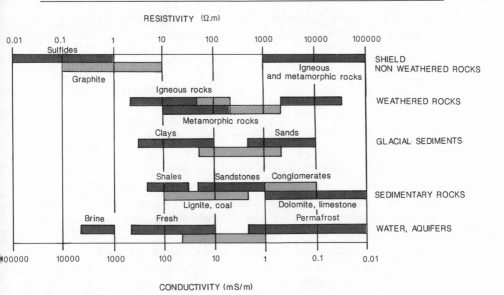

Fig. VIII.2. Range of resistivity variations for terrestrial materials. (After Palacky, G. J., 1987, Resistivity Characteristics of Geologic Targets, in *Electromagnetic Methods in Applied Geophysics*, vol. 1, ed. M. N. Nabighian, Series: Investigations in Geophysics, vol. 3, Soc. of Expl. Geophys., Tulsa, Okla.)

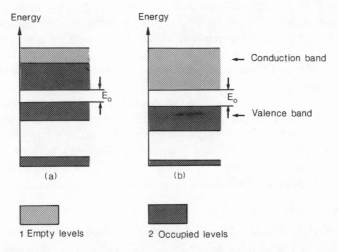

Fig. VIII.3. Two possible cases of occupation of the indicated electron energy states in solids. (*a*) Metal: the conduction band is partially filled. (*b*) Insulator: the conduction band is totally empty (Semi-conductor: $E_o < 2.5$ eV; Dielectric: $E_o > 2.5$ eV).

Table VIII.1. ENERGY GAPS BETWEEN VALENCE AND CONDUCTION BANDS

Crystal	E_g (eV)
Si	1.14
Ge	0.67
Diamond	5.33
Olivine	0.5

empty band by thermal energy alone. Semiconductors are the intermediate case: they have completely full energy bands, but the energy gap is
sufficiently small. At moderate temperatures a small number n of
electrons are thermally excited into the conduction band: n varies with
temperature according to the law $n \sim \exp(-E_g/2kT)$, where k is
Boltzmann's constant and T is the absolute temperature. If $E_g \gg kT$,
the number of electrons excited into the conduction band is negligibly
small. Table VIII.1 shows the energy band gap for several minerals.
Note that the band gap for Si is slightly more than twice that of Ge.
This implies that everything else being equal, Si would require twice as
large a temperature as Ge to have the same number of carriers as Ge.
Because kT is equal to 1 eV at a temperature $T = 11{,}000°K$, the
number of conduction electrons is totally negligible for a solid like
diamond. The electronic conductivity varies enormously from solid to
solid due to the variability in E_g, the exponential dependence $n \sim$
$\exp(-E_g/2kT)$, and the relation VIII.4. For insulators $n \approx 0$ and for
conductors such as copper it is $n \approx 1$ per atom.

The mobility of the electrons is the second important parameter
entering equation VIII.4. Electron mobility is controlled by its collisions
with the lattice. If τ is the time between two collisions, m and e the
mass and charge of an electron, then $m\vec{v} = q\vec{E}\tau$ is the momentum
acquired by an electron between collisions. The electron mobility is thus
$\mu = q\tau/m$. As the temperature T increases, the frequency of collisions
increases and τ decreases: thus μ varies with T, but its dependence on
T is not as great as that of n.

Ionic Conduction

Most of the minerals discussed previously are dielectrics (insulators)
under the temperature and pressure conditions of the earth's crust. In
dielectrics the important charge carriers are ions rather than electrons.
The contrast in conductivity between conductors and insulators results
from two causes: there are fewer ionic charge carriers in dielectrics than
there are conduction electrons in metals, and due to their larger volume

and mass the ionic mobilities are much smaller. Ionic displacement occurs primarily by diffusion. In chapter IV we had considered such displacement processes as occurring by diffusion or by dislocation motion. In the same way that a diffusional flux results from the migration of atoms (or vacancies) in response to a chemical potential gradient, ionic electrical conduction results from the migration of ions in response to an electrical potential gradient. In the general case the flux of particles (ions, atoms, etc.) is the result of diffusion in response to a concentration gradient and transport in response to an applied force:

$$J = -D\frac{\partial n}{\partial x} + \bar{v}n.$$

The term $D\frac{\partial n}{\partial x}$ is the flux due to diffusion (Fick's law) where D is the diffusion coefficient, and the term $\bar{v}n$ is due to convective transport ($\bar{v}$ being the average particle velocity). In a stationary regime, the total flux J is zero, therefore:

$$D\frac{\partial n}{\partial x} = \bar{v}n.$$

Now suppose that at thermodynamic equilibrium, the spatial distribution $n(x)$ of particles follows Boltzmann's law:

$$n(x) = n_o\exp\left(-\frac{\varphi}{kT}\right)$$

where the potential energy φ shown here can be either the chemical potential (cf. chap. IV) or the electrical potential qV. This results in

$$\frac{dn}{dx} = -\frac{n}{kT}\frac{d\varphi}{dx} = \frac{nF}{kT}$$

with $F = -\frac{d\varphi}{dx}$ being the applied force responsible for the convective transport. Combining this result with the relation $D\frac{dn}{dx} = n\bar{v}$, one obtains the Nernst-Einstein equation

$$\frac{\bar{v}}{D} = \frac{F}{kT}.$$

Equation IV.24 corresponded to the case where the potential φ was the chemical potential. In the present case $F = qE$, the potential is an electrical potential. The ionic mobility is thus

$$\mu = \frac{qD}{kT}. \qquad\qquad \text{VIII.5}$$

Recalling that the current flux is given by $J = nq\bar{v} = nq^2\dfrac{D}{kT}E$, the conductivity becomes

$$\sigma = nq^2\frac{D}{kT}.$$ VIII.6

The diffusion coefficient is often found to follow an exponential law $D = D_o\exp(-\dfrac{E_o}{kT})$, where E_o is the activation energy. The density of carriers n also obeys an exponential law. Thus the density of carriers and mobility are in this case two similar functions of T.

Besides the intrinsic ionic conduction just discussed, there can exist an extrinsic conduction due to the presence of impurities. These impurities can be a different species of mobile ions or they can be ions which contribute additional conduction electrons. The total conductivity is then the sum of two processes (intrinsic and extrinsic):

$$\sigma = \sigma_{io}\exp\left(-\frac{E_i}{kT}\right) + \sigma_{eo}\exp\left(-\frac{E_e}{kT}\right).$$

At lower temperatures, the dominant process is the one with the lower activation energy, which in general is the extrinsic conductivity σ_e. At higher temperature, the term with the higher activation energy can become dominant due to a much larger coefficient σ_{eo}. This is well illustrated by the mineral olivine. It has both an intrinsic electronic conductivity ($E_g \sim .5$ eV and $\sigma_0 \sim 10^2$ S/m) and an intrinsic ionic conductivity ($E_i \sim 3$ eV and $\sigma_o \sim 10^8$ S/m): the ionic conductivity becomes increasingly more important at higher temperatures. Table VIII.2 presents the activation energies and preexponential factors for several important rocks.

Table VIII.2. PARAMETERS DEFINING THE TEMPERATURE DEPENDENCE OF CONDUCTIVITY

Rock	E(electronic) (eV)	E(ionic) (eV)	σ_o(electronic) (S / m)	σ_o(ionic) (S / m)
Granite	0.62	2.5	5×10^{-4}	10^5
Gabbro	0.70	2.2	7×10^{-3}	10^5
Basalt	0.57	2.0	7×10^{-3}	10^5
Peridotite	0.81	2.3	4×10^{-2}	10^5
Andesite	0.7	1.6	6×10^{-3}	

Source: After Keller, G. V., 1987, Rock and Mineral Properties, in *Electromagnetic Methods in Applied Geophysics*, vol. 1, edited by M. N. Nabighian, Series: Investigations in Geophysics, vol. 3 (Soc. of Expl. Geophys, Tulsa, Okla).

Conduction in Solutions

Waters which are present in the pore space of a rock often contain a wide variety of salts in solution. The nature of the salts reflects on the precise origin and history of the water. For example, connate (fossil) water is water left over from the time of deposition and is typically rich in NaCl. Salts such as NaCl and KCl dissociate totally in solution and can move independently under the influence of an applied electric field. The movement of anions and cations in opposite directions result in an electric current. At steady state, the ion velocity $\vec{v}$ is given by Stoke's law: the applied force, $q\vec{E}$, is balanced by the viscous force, $6\pi\eta r \vec{v}$, exerted by the fluid on the hydrated ion.

$$q\vec{E} = 6\pi\eta r \vec{v}.$$

Here η is the viscosity of the water and r is the effective radius of the ion. The ion mobility is $\mu = \dfrac{q}{6\pi\eta r}$, therefore the ionic conductivity $\sigma_w = qn\mu$ (equ. VIII.4) becomes

$$\sigma_w = \frac{nq^2}{6\pi\eta r}. \qquad \text{VIII.7}$$

The conductivity is proportional to the ionic concentration n (fig. VIII.4a). At fixed concentration there is a variation of approximately

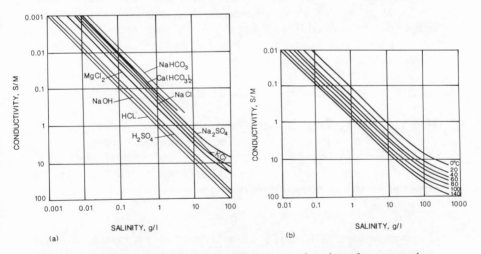

Fig. VIII.4. Conductivity variations of solutions as a function of concentration. (*a*) For different salts in solution (at 20°C). (*b*) For a solution of NaCl at different temperatures. (After Keller, G. V., 1987, Rock and Mineral Properties, in *Electromagnetic Methods in Applied Geophysics*, vol. 1, ed. M. N. Nabighian, Series: Investigations in Geophysics, vol. 3, Soc. of Expl. Geophys., Tulsa, Okla.)

one order of magnitude between the various salts due to their relative mobilities. The viscosity of water is controlled by the hydrogen bonding between water molecules and is a function of temperature. Near 0°C, the viscosity decreases with increasing temperature. At much higher temperatures (400°–800°C), the viscosity increases with T (fig. VIII.4b). At fixed concentration, σ_w is approximately one order of magnitude higher at 0°C than at 140°C.

CONDUCTIVITY OF ROCKS

The effective conductivity of an aggregate depends in a very sensitive fashion on its microstructure. This results from the enormous contrast in conductivity between insulating silicates ($\sigma \sim 10^{-14}$ to $10^{-10} Sm^{-1}$) and saline solutions ($\sigma \sim 10^{-1}$ to $1\ Sm^{-1}$), which implies that electrical currents will flow almost totally through the saline pore water within the rock. Pore microstructure thus dictates the path of the electric currents.

Effective Conductivity: Definition and Limiting Values

In the interior of a rock, Ohm's law applies locally:

$$\vec{J}(r) = \sigma(r)\vec{E}(r) = -\sigma(r)\vec{\nabla}V(r).$$

Due to the pore scale variations in conductivity $\sigma(r)$, the potential at the pore scale, $V(r)$, will depend on the microgeometry. The effective conductivity, σ_{eff}, of a macroscopically homogeneous and isotropic sample can be defined by applying a potential difference ΔV between two parallel planes located at $z = 0$ and L and measuring the average current density $\langle J \rangle$ parallel to the applied field:

$$\langle \vec{J} \rangle \equiv \frac{1}{A} \iint_{S} \left(\vec{J} \cdot \hat{z} \right) dS \equiv \sigma_{\text{eff}} \left(\frac{\Delta V}{L} \right).$$

Here A is the cross-sectional area perpendicular to $\hat{z}$, the unit vector in the z-direction. σ_{eff} (as all other effective properties) depends on the volume fractions of the rock constituents, their individual conductivities, and the rock microstructure. For a time-independent electric field, the equation for charge conservation becomes

$$\vec{\nabla} \cdot \vec{J}(r) = -\vec{\nabla} \cdot \left(\sigma(r)\vec{\nabla}V(r) \right) = 0.$$

If the microscopic distribution of conductivity $\sigma(r)$ is known, we could in principle solve for $V(r)$, and use the defining relation for $\langle J \rangle$ to compute σ_{eff}

$$\sigma_{\text{eff}} \equiv \left(\frac{L}{\Delta V} \right) \frac{1}{A} \iint_{S} \sigma(r)(-\hat{z} \cdot \vec{\nabla}V)\, dS. \qquad \text{VIII.8}$$

Equation VIII.8 relates the conductivity at the sample scale, σ_{eff}, in terms of the microscopic conductivity at the pore scale, $\sigma(r)$. It is only possible to directly calculate σ_{eff}, for the simple cases such as those discussed in chapter III. If direct computation is impractical, it is still possible to obtain *upper and lower bounds* on the effective conductivity (cf. equ. III.5). Suppose for example that a rock is composed of N minerals of conductivities $\{\sigma_1 \ldots \sigma_i \ldots \sigma_N\}$ and volume fractions $\{x_1 \ldots x_i \ldots x_N\}$. If the minerals are arranged in a layered structure perpendicular to the applied field, the current is constant in each layer $\vec{J} = \sigma_i(\overrightarrow{\nabla V})_i$ and $x_i = L_i/L$. Then equation VIII.8 yields:

$$\frac{1}{\sigma_{\text{eff}}} = \sum_{i=1}^{N} \frac{x_i}{\sigma_i} \qquad \text{VIII.9}$$

the classical result for resistances in series. If the field is applied parallel to the layers, the potential difference is constant $(\overrightarrow{\nabla V})_i = \dfrac{\nabla V}{L}$ for every layer i. The classical result for resistances in parallel is recovered.

$$\sigma_{\text{eff}} = \sum_{i=1}^{N} x_i \sigma_i. \qquad \text{VIII.10}$$

We have seen in chapter III that the above relations provide upper and lower bounds for σ_{eff}.

The preceding bounds are not very limiting because they do not use any other information except the values $\{x_i\}$ and $\{\sigma_i\}$. It is possible, using supplementary information, to further restrict the domain of possible values for σ_{eff}. The bounds of Hashin-Shtrikman introduced in chapter III are applicable for an isotropic porous media. The equivalent result to III.4 for the conductivity is:

$$\sigma^* = \sigma_1 + \frac{v_2}{\dfrac{1}{\sigma_2 - \sigma_1} + \dfrac{v_1}{3\sigma_1}} \qquad \text{VIII.11}$$

For the case where the two constituents are water of conductivity σ_w and solid rock of conductivity σ_s, one obtains with $\delta\sigma = \sigma_w - \sigma_s$:

$$\sigma'' \le \sigma_{\text{eff}} \le \sigma'$$

with:

$$\sigma'' = \sigma_s \left[1 + \frac{3\phi\,\delta\sigma}{3\sigma_s + (1 - \phi)\,\delta\sigma} \right]$$

and

$$\sigma' = \sigma_w \left[1 - \frac{3(1 - \phi)\,\delta\sigma}{3\sigma_w - \phi\,\delta\sigma} \right]$$

where ϕ is the porosity.

The upper bound is attained for a microstructure of spherical solid grains, each coated with a shell of water in the ratio $1 - \phi$ to ϕ. In order to fill all space the coated spheres must have a distribution of sizes, ranging to the infinitesimal. For this microstructure the fluid forms a percolating cluster, while the solid grains are isolated from each other. The lower bound corresponds to an identical microstructure with the two components interchanged. In this case the solid phase forms the percolating cluster, while the pores are isolated. In the limit of non-conducting grains, $\sigma_s = 0$, the lower bound is always zero, while the upper bound is $\sigma_w \left(\dfrac{2\phi}{3 - \phi} \right)$. The H-S bounds are quite useful unless the ratio σ_w / σ_s becomes too great as can be seen in figure VIII.5. For most saturated rocks the bounds are insufficient to constrain the electrical conductivity and more elaborate models are necessary.

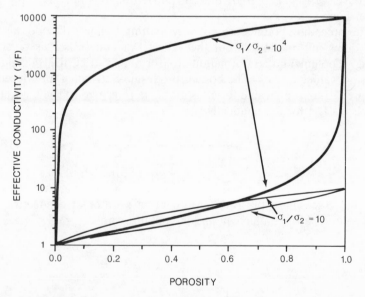

Fig. VIII.5. Hashin Shtrikman bounds for a two-component medium whose ratio of conductivities are $\sigma_1 / \sigma_2 = 10$ and 10^4.

Formation Factor

The differences in conductivity between the mineral grains comprising the rock are usually very small as compared to the differences in conductivity between the minerals and water. A fluid saturated rock can therefore be modeled as a two-component medium: solid grains (volume fraction $1 - \phi$ and conductivity σ_s), saline water (volume fraction ϕ and conductivity σ_w), with their ratio being $\sigma_s/\sigma_w \approx 10^{-10}$. The effective conductivity of a rock saturated with a fluid will be proportional to the conductivity of the fluid σ_w, but reduced by a factor $1/F$:

$$\sigma_{\text{eff}} \equiv \frac{\sigma_w}{F}. \qquad \text{VIII.12}$$

The factor F, commonly called the formation factor, depends on the rock microstructure and the ratio σ_s/σ_w. Because the ratio σ_s/σ_w is essentially zero for most rocks, the formation factor is only a function of the porosity microstructure. As $\phi \to 1$, $F \to 1$, and $\sigma_{\text{eff}} \to \sigma_w$. At the other extreme as $\phi \to 0$, $F \to \sigma_w/\sigma_s \approx 10^{10}$, and $\sigma_{\text{eff}} \to \sigma_s$.

Note that if the grains are assumed to be non-conducting, $\sigma_s = 0$, and the conductivity of the water is uniform, $\sigma_w = $ constant, then equation VIII.8 reduces to the result:

$$\frac{1}{F} = \frac{\sigma_{\text{eff}}}{\sigma_w} = \left(\frac{L}{\Delta V}\right)\frac{1}{A}\iint\limits_{S_p}(-\hat{z}\cdot\vec{\nabla}V)\,dS. \qquad \text{VIII.13}$$

The integration is only over the pore area S_p of the sample cross-section A, $S_p \subset A$. For an isotropic media $\phi = S_p/A$. The formation factor can be viewed as a reduction in brine conductivity, due to the presence of an insulating phase. Part of the reduction is due to the fact that only a fraction ϕ is conducting, the other part is due to effects of tortuosity. One defines the tortuosity α through the relation

$$\frac{1}{F} = \frac{\phi}{\alpha}. \qquad \text{VIII.14}$$

α takes into account effects such as: the lack of pore space connectivity, their tortuous paths around the solid grains, and variable pore cross-sections. Equation VIII.13 implies that dead-end or isolated pores do not contribute to the conductivity because they would have very small or zero gradients. The integral accentuates well-connected pores that have large gradients (i.e., small pore throats). For capillary tubes of uniform cross-section and oriented parallel to the field, the gradient of V would be constant and give the expected result $\alpha = 1$. If these same capillary tubes had a variable cross-section, $\alpha > 1$.

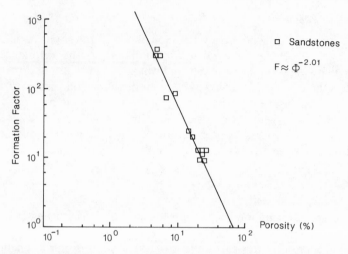

Fig. VIII.6. Empirical relation (Archie's law) between formation factor and porosity for the Vosges and Fontainebleau sandstones. (After Ruffet, C., Y. Guéguen, and M. Darot, 1991, Complex Conductivity Measurements and Fractal Nature of Porosity, *Geophysics* 56:6758–68.)

Archie's Law

In the absence of detailed information on pore microstructure, one must rely on empirical relations between the formation factor and porosity, the most widely used being Archie's law. These empirical relationships are of particular importance in the petroleum industry because resistivity measurements serve as one of the principal methods for estimating fluid saturations in oil-bearing rocks. Archie's law relates the porosity and formation factor for a rock completely saturated with water:

$$F = \phi^{-m} \qquad\qquad\qquad \text{VIII.15}$$

or

$$F = (\phi - \phi_o)^{-m}. \qquad\qquad\qquad \text{VIII.16}$$

The exponent m is approximately constant for a given rock type. For most sedimentary rocks it is found to be in the range $1.5 \le m \le 2.5$, with the majority of sandstones being close to $m = 2$ (fig. VIII.6). Because data can be represented easier with two parameters, many investigators have also used Archie's law in the form $F = b\phi^{-m}$. This has been found to be necessary when conductivity data for rocks with variable diagenetic histories are analyzed simultaneously. Although equations VIII.15 and 16 are referred to as "laws," they should be viewed as empirical relations which fit sedimentary rock data over a limited porosity range, $.05 < \phi < .40$. There is no theoretical reason why this functional form should fit all microstructures for all values of porosity. We can make the same remark here as we made in chapter V when discussing relations between permeability and porosity: the existence of correlations $k \sim \phi$ or $F \sim \phi$ is determined by the rock history, through the evolution of its microstructure.

In the form VIII.15 Archie's law implies that the porous network forms an infinite percolating cluster, in the sense of percolation theory. On the other hand, equation VIII.16 suggests the existence of a percolation threshold at $\phi = \phi_o$. We shall examine later how percolation theory can lead to results which approach Archie's law.

Oil and Gas Saturation in Rocks

Until now we have only considered rocks that are 100% saturated by an aqueous electrolyte conductor (saline water). However, the pore space of a rock is not always completely saturated by a conducting fluid: sedimentary rocks above the water table may be partially saturated by air, while reservoir rocks are partially saturated by oil. Due to the very low conductivity of oil, natural gas, and air as compared to water, increasing the saturation of these fluids will significantly decrease the bulk rock conductivity.

Determining the saturation S_o (oil) and S_w (water) is one of the most economically important rock physics problems. The product of oil saturation and porosity, $S_o \phi$, is proportional to the volume of oil (or gas) that exists in a reservoir. The relative permeability of oil, which depends on S_o, indicates how rapidly the oil can be extracted. Porosity and permeability are intrinsic rock properties. This is not the case for saturation and relative permeability, which depend on how the oil and water are distributed through the pore space. How two immiscible fluids are distributed depends in detail on the dynamical evolution of the system and wetting characteristics of the fluids. Consider the patterns that can develop when an invading fluid (oil) displaces an initial fluid (water) or vice versa. If the invading fluid is the wetting fluid, it will first fill the smallest pores. If the initial fluid is the wetting fluid, then the invading fluid will first fill the accessible large pores. Invasion by a non-wetting fluid is controlled by capillary forces and can be simulated numerically with a network model. As pressure is increased the fluid advances pore by pore, into the most favorable pores (those with the largest entry radius). Such a process guarantees that the invading fluid always forms a connected cluster. It also allows the possibility that the initial fluid can be bypassed and be completely enclosed by the invader (become an isolated cluster). By injecting air into paraffin oil on a

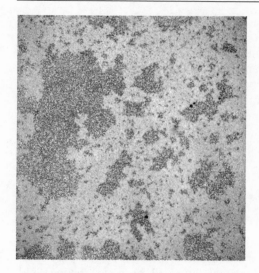

Fig. VIII.7. Injection of air (non-wetting fluid) into paraffin (wetting fluid) on a two-dimensional (350 × 350) random network. At low injection pressures, the oil is displaced pore by pore by the process of invasion percolation. The light zones correspond to air, while the dark zones correspond to oil. Isolated clusters of oil can be seen. Air is injected from the left edge. A semi-permeable membrane prevents the non-wetting fluid from leaving at the right. (After Lenormand, R., and C. Zarcone, 1985, Invasion percolation in an Etched Network: Measurement of a fractal Dimension, *Phys. Rev. Lett.* 54:2226–29.)

two-dimensional random network, Lenormand (1985) experimentally demonstrated this process (fig. VIII.7).

To predict the saturation of a rock a dimensionless parameter called the resistivity index (RI) is utilized.

$$RI = \frac{\rho_t}{\rho_o}.$$ VIII.17

RI is the ratio of the resistivity of a partially saturated rock, ρ_t, to the resistivity when it is completely saturated with a conducting fluid, ρ_o. RI is a function of S_w and the pore microstructure. Archie's second law, also empirically determined, is expressed by the relation:

$$RI = (S_w)^{-n}$$ VIII.18

where n, the saturation exponent, is approximately constant for a given porous medium and a given system of fluids. Equation VIII.18 can be justified through the following argument. Assume that the non-conductive fluid displaces the water in a homogeneous fashion: the same fraction of water is displaced from all pores independent of their size. The remaining volume fraction of water would be ϕS_w, and according

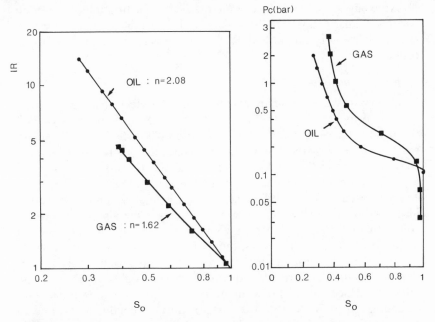

Fig. VIII.8. (*a*) Resistivity index versus saturation ($RI = S_o^{-n}$) for the same sandstone saturated with oil or gas (non-wetting fluids) and water (wetting fluid). (*b*) Drainage curves for the two preceding cases [$k_o = 40$ md, $\phi = 0.18$]. (After Longeron, D. G., M. J. Argaud, and J.-P. Feraud, 1986, Effect of Overburden Pressure, Nature, and Microscopic Distribution of the Fluids on Electrical Properties of Rock Samples, Soc. Petrol. Eng. paper 15383.)

to Archie's law, the resistivity of the rock would be $\rho_t = (\phi S_w)^{-m}\rho_w$. The resistivity index then becomes

$$RI = \frac{\rho_t}{\rho_o} = \frac{(\phi S_w)^{-m}\rho_w}{\phi^{-m}\rho_w} = (S_w)^{-m}.$$

In reality, because of the variation in pore and pore throat sizes as well as the inhomogeneous distribution of microporosity, oil does not displace water uniformly throughout the rock. In addition, water is not necessarily the wetting fluid, thus the distribution of oil after drainage or imbibition is quite different (cf. chap. II). It is thus not surprising that the exponent *n* is usually found to be different from *m* as is illustrated in figure VIII.8. The difference between drainage curves is due to surface tension differences ($\gamma = 27.4$ mN/m for oil and $\gamma = 59$ mN/m for gas). At equal saturations the distribution of fluids, and thus exponent *n*, are different. A value of *n* determined from a gas-brine drainage curve, will be an inaccurate predictor of saturations in oil-brine systems. Thus to apply the resistivity index technique, it is necessary that the laboratory-established RI-S_o relation is representative of in situ conditions.

Surface Conductivity

Until now we have implicitly assumed that the electrolyte conductivity was uniform and that the mobile ions were uniformly distributed throughout the pore space. This assumption is satisfied by sedimentary rocks such as clay-free sands. But in many rocks there are additional charge carriers in the fluid phase, adjacent to solid surfaces, giving rise to an additional component of conduction along the surface. The most common occurrence of surface conduction is in shaly sands, where the insulating sand grains are overgrown by clay minerals whose surface conductivity cannot be neglected. Clay minerals usually contain charged impurities, such as Al^{3+} substituting for SI^{4+}. The charge deficit is compensated by cation adsorption on the mineral surface. This adsorption results in the formation of an electrical double layer near the mineral surface (fig. II.15) with the concentration of ions in solution extending 30 to 80 Å from the surface. Surface conduction is due to these ions in the diffuse double layer. When the pore space is fully saturated with an electrolyte and a field applied, the diffuse layer ions are displaced along the interfaces. Because the layer thickness is very small compared to the pore diameter (sandstone pores > 0.1 μm in general), surface conduction is localized near the clay surfaces and is proportional to the surface area. Formally, the conductivity $\sigma(r)$ can be written as the sum of the normal ionic brine conductivity σ_w and a near-surface term $\delta\sigma(r)$ due to the double layer, that is, $\sigma(r) = \sigma_w + \delta\sigma(r)$. $\delta\sigma(r)$ is only non-zero within the double layer. Equation VIII.13 then leads to the result

$$\sigma_{\text{eff}} = \frac{1}{F}\left(\sigma_w + \frac{2\sigma_{\text{surf}}}{\Lambda}\right) \equiv \frac{1}{F}(\sigma_w + BQ_v)$$

where F is the formation factor with no surface conduction present. $\Lambda/2$ is a characteristic length of the rock microstructure and is approximately equal to the ratio (pore volume)/(pore surface). For the special case of a cylindrical capillary, Λ would be the radius. Surface conductivity, σ_{surf}, is proportional to the number of ions per unit surface area, n', their charge q, and their mobility B: $\sigma_{\text{surf}} = n'qB$. With $\Lambda/2 = V_p/S_p$, one obtains $\dfrac{2\sigma_{surf}}{\Lambda} = B\dfrac{n'qS_p}{V_p} = BQ_v$. Q_v is the total charge in the double layer per unit pore volume.

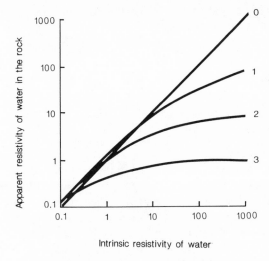

Fig. VIII.9. Surface conductivity effects due to the presence of clays. (After Keller, G. V., 1987, Rock and Mineral Properties, in *Electromagnetic Methods in Applied Geophysics*, vol. 1, ed. M. N. Nabighian, Series: Investigations in Geophysics, vol. 3, Soc. of Expl. Geophys., Tulsa, Okla.) 0: no clays; 1: few clays; 2: small amount of clays; 3: significant amount of clays.

The previous equation is valid when the formation factor F is the same for both types of ions (double layer and bulk electrolyte) and when $\sigma_w \gg BQ_v$. For very pure water, σ_w is very small and the term BQ_v becomes comparable in size. This is illustrated in figure VIII.9 where the resistivity $1/\sigma_w$ of the bulk solution is plotted against the apparent resistivity of the water in the pore space $1/(\sigma_w + BQ_v)$. The straight line represents no clays being present, $BQ_v = 0$. When various amounts of clay are added, $BQ_v \neq 0$, the water resistivity is smaller than the bulk resistivity due to surface conduction, with the effect being larger for high-resistivity water. Waxman and Smits (1968) generalized the previous equation by assuming that the surface conduction will be in parallel with the bulk conduction for all values of the bulk conductivity, and that the formation factor F should be replaced by the factor F^*.

$$\sigma_{\text{eff}} = \frac{1}{F^*}[\sigma_w + BQ_v].
\qquad \text{VIII.19}$$

The factor F^* is determined from the slope of the curves in figure VIII.9 in the low resistivity limit, $\sigma_w \gg BQ_v$.

Role of Conducting Minerals

The conductive properties of most rocks can essentially be attributed to the presence of water. But, certain rocks which contain high concentrations of conductive minerals such as sulfides, magnetite, graphite, pyrite, and carbon also can have significantly higher conductivities even in the absence of water. The resistivities of these minerals (table VIII.3) are much smaller than those of silicates, which average around 10^{12} ohm-m at room temperature.

Table VIII.3. THE RESISTIVITY OF COMMON CONDUCTIVE MINERALS

Mineral	Resistivity (ohm-m)
Graphite (parallel to cleavage)	$36-100 \times 10^{-8}$
Graphite (perpendicular to cleavage)	$28-9900 \times 10^{-8}$
Galena (PbS)	$6.8 \times 10^{-6}-9.0 \times 10^{-2}$
Pyrite (FeS$_2$)	$1.2-600 \times 10^{-3}$
Magnetite (Fe$_3$O$_4$)	52×10^{-6}

The distribution of conducting minerals is an important factor in determining the overall rock conductivity. For example, rather small concentrations of graphite will significantly increase rock conductivity due to its tendency to form thin continuous films along grain boundaries. Many minerals which are known to render rocks conductive occur in dendritic patterns. These patterns often arise after hydrothermal transport and mineral deposition along connected rock pathways. Other conductive minerals occur primarily as randomly isolated grains which may be enveloped in a coating of insulating minerals as calcite or quartz. In these cases the volume fraction must be significantly larger before a continuous percolating structure can exist. Figure VIII.10

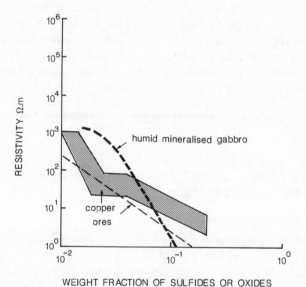

Fig. VIII.10. Variation of resistivity with the amount of conducting minerals. (After Keller, G. V., 1987, Rock and Mineral Properties, in *Electromagnetic Methods in Applied Geophysics*, vol. 1, ed. M. N. Nabighian, Series: Investigations in Geophysics, vol. 3, Soc. of Expl. Geophys., Tulsa, Okla.)

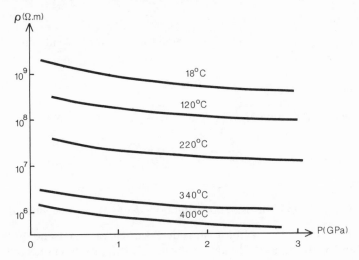

Fig. VIII.11. Variation of resistivity with pressure for a basalt. (After Parkhomenko, E. I., 1967, *Electrical Properties of Rocks*, translated from Russian and edited by G. V. Keller, New York: Plenum Press.

shows the observed relationship between rock resistivity and concentration of conducting minerals. The main feature of this data is that there can be a dramatic decrease in resistivity when the volume fraction of conductive minerals is greater than 10%. Variations of this magnitude are easily detectable by electromagnetic prospecting.

Variation with Pressure

The effects of pressure on the conductivity of insulators and semiconductors, although quite small, are measurable. As the pressure increases, the electron band gap E_o decreases as the conduction and valence bands approach each other and the electronic conductivity increases. Ionic conductivity in general varies inversely with pressure. Figure VIII.11 shows however that pressure effects are small as compared to temperature. Considering typical values of the geothermal gradient (30°C/km) and pressure gradient (270 bars/km) near the earth's surface, one sees that in the leading approximation pressure effects can be neglected.

CONDUCTIVITY MODELS

The various theoretical models of rock conductivity are very similar to those for permeability, when the analogy is made between the flow of electric current and the circulation of conducting fluids.

Statistical Models

The tubes and fissures model examined in chapter V leads directly to the relation

$$F \cong \frac{1}{f\phi}.$$
VIII.20

The electrical current traversing a cross-section of area S is:

$$\vec{J} = \sigma \vec{E} S = f\sigma_w \vec{E} \sum_{i=1}^{N} s_i$$

where the s_i are the intersections of S with the different conductors (cf. figs. V.1 and 2). With the approximation $\phi S = \sum_{i=1}^{N} s_i$, one immediately obtains VIII.20. The factor f, introduced in chapters III and V, is the fraction of the random conductors which form an interconnected network. The results of chapter V showed that $f \sim \dfrac{1}{\tau^2}$, where τ is the hydraulic tortuosity. Comparing VIII.14 and VIII.20 shows that $f \sim \dfrac{1}{\alpha}$ where α is the electrical tortuosity introduced previously. When $f \to 1$, $\alpha \to 1$: the network is perfectly connected.

Kirkpatrick's Model

Kirkpatrick (1973) described how the effective conductivity of an inhomogeneous medium can be obtained from the microscopic electrostatic equations:

$$\vec{J}(r) = -\sigma(r)\vec{\nabla}V(r)$$

$$\vec{\nabla} \cdot \vec{J}(r) = 0.$$

By introducing a finite difference scheme on a grid of points $\{r_i\}$, the vertices of a cubic lattice of bond length Δr, he obtained the following system of linear equations for the voltages $V_i \equiv V(r_i)$.

$$\sum_{j} g_{ij}(V_i - V_j) = 0 \quad \text{with } g_{ij} = \sigma\left(\frac{r_i + r_j}{2}\right).$$

Each site i in the interior of the lattice has z nearest neighbor sites j. This set of equations is identical to those of Kirchoff for a network of conductances g_{ij}. Thus the real medium is modeled as a network of variable conductances g_{ij} (fig. VIII.12a). For example, the conductance of a cylinder of radius R and length L which is saturated with a conductive liquid is $g = \sigma_w \dfrac{\pi R^2}{L}$. This random conductance network is

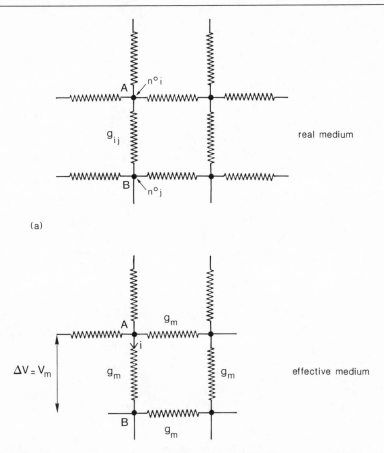

Fig. VIII.12. (*a*) A real medium modeled by a network with a distribution of bond conductances g_{ij}. (*b*) An equivalent medium with all conductances equal to g_m.

equivalent to a network of the same topology which has identical conductances g_m (fig. VIII.12b). The effective conductance of the cubic network is thus $g_{\text{eff}} = g_m$, where g_m is determined through the following criteria:

if one of the conductances g_m of the effective (homogeneous) medium is replaced by one of the local conductances $g_{ij} = g_o$, the change in potential δV created between A and B, when averaged over the whole ensemble of random conductances g_{ij}, will be zero: $\langle \delta V \rangle = 0$.

Let i be the current between A and B (fig. VIII.12b): $i = g_m V_m$, if V_m is the difference in potential between two adjacent nodes. By replacing

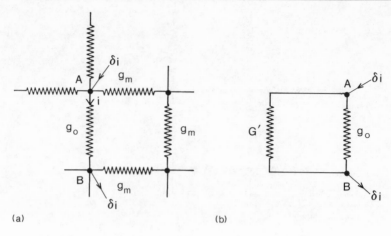

(a)   (b)

Fig. VIII.13. (a) A conductance g_m of the effective medium is replaced by a value $g_{ij} = g_o$. The perturbation of the potential is compensated by δi. (b) δi is obtained by considering the equivalent conductance ($G' + g_o$).

g_m by g_o (fig. VIII.13a) a perturbation in potential δV is introduced: $V_a - V_b \neq V_m$. To reestablish this equality, imagine introducing a fictitious current δi at A and extracting it at B. When reestabilished, the total current between A and B must be $i' = g_o V_m$. the difference $i - i'$ is:

$$\delta i = i - i' = (g_m - g_o)V_m.$$

The perturbation in potential is thus

$$\delta V = \frac{\delta i}{g_o + G'}$$

where G' is the equivalent conductance between A and B in the absence of g_o (fig. VIII.13b): $G' = G - g_m$ if G is the equivalent total conductance between A and B for the network shown in 12b. Thus:

$$\delta V = \frac{\delta i}{g_o + G - g_m} = \frac{g_m - g_o}{g_o - g_m + G}V_m.$$

The equivalent conductance G for the effective medium is obtained in the following manner:

(a) a current i_o is introduced at A and extracted at very distant points in all directions.

(b) a current i_o is introduced equally at distant points and extracted at B.

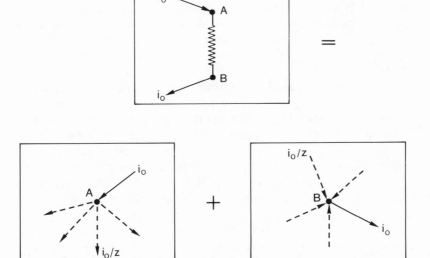

Fig. VIII.14. Distribution of the equivalent current i_o introduced at A and extracted at B: $i_o = GV_{AB}$.

Superposition of these two steps (fig. VIII.14) implies that as i_o is introduced at A, due to symmetry, i_o/z flows in each bond away from A. When i_o is extracted at B, again due to symmetry, i_o/z flows in each equivalent bond to B. Thus a total $2i_o/z$ flows from A to B:

$$V_{AB} = \frac{2i_o}{z}\frac{1}{g_m} = \frac{i_o}{G} \quad \text{or} \quad G = \frac{z}{2}g_m.$$

When this result is combined with the previous equation, one obtains

$$\delta V = \frac{g_m - g_o}{g_o + \left(\dfrac{z}{2} - 1\right)g_m}V_m.$$

The criterion determining g_m is $\langle \delta V \rangle = 0$. When there is a distribution $f(g)$ in the conductances, either continuous or discrete, g_o takes on all the values g.

$$\left\langle \frac{g_m - g}{g + ((z/2) - 1)g_m} \right\rangle = 0. \qquad \text{VIII.21}$$

Equation VIII.21 is an accurate approximation for g_m for most distributions of conductances. But when the distribution has a high density of low or zero conductances and percolation effects are important, VIII.21 no longer gives a good estimate for the effective conductance. In those cases more precise results can be obtained from the numerical simulations over a network (David et al., 1990).

Percolation Models

Because the pore space of sedimentary rocks is sufficiently complex and varied, it can be viewed as a random network with a very broad distribution of conductances. Transport in such a network is dominated by conductances with magnitudes greater than a characteristic value g_c. The value of g_c is such that the set of conductances $\{g\}$ which are greater than g_c, $\{g|g > g_c\}$, forms an "infinite" connected cluster in the sense of percolation theory. From this point of view the network conductivity is associated with a percolation problem whose threshold value is g_c. This approach was first used by Ambegaokar et al., (1971) for the electrical conductivity of amorphous semiconductors. Ambegaokar's method furnishes a conceptual basis for many percolation models of transport in rocks, in particular the model of Katz and Thompson (1987) which models porosity as a cluster of tubes, and the model of Charlaix et al. (1987) which assumes the porosity is due to fractures. Why this conceptually simple but elegant approach should work is based on the following reasoning. If we place the bonds on the network starting with the largest, then when the bond of conductance g_c is placed into a network, it completes a connected cluster which spans the network, with all other bonds in that cluster being *larger* than g_c (figs. VIII.15 and 16). This implies that the bond g_c is in series with the

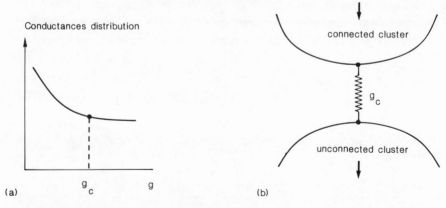

Fig. VIII.15. (*a*) Ambegaokar's model: g_c is the "critical" conductance which completes the infinite cluster spanning the network. (*b*) Construction of the infinite cluster: g_c is the final link.

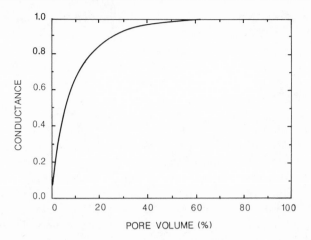

Fig. VIII.16. Conductance of a sandstone as a function of the pore fraction occupied by mercury. (After Thompson, A. H., A. J. Katz, and C. E. Krohn, 1987, The Microgeometry and Transport Properties of Sedimentary Rocks, *Advances in Physics* 36, n. 5, pp. 625-94.)

equivalent conductance g_o of the other bonds in this cluster. The effective conductance g_{eff} of the percolating cluster is then

$$\frac{1}{g_{\text{eff}}} = \frac{1}{g_c} + \frac{1}{g_o}.$$

Because g_o is composed of many bonds, all conductances being greater than g_c, it is certain that $g_o \gg g_c$. Therefore $g_{\text{eff}} \approx g_c$. The remaining network is composed of bonds with conductances less than g_c that are essentially in parallel. Their contribution is also quite small. From this viewpoint we can conclude that the network conductivity is dominated by a critical subnetwork composed of bonds with conductances g greater than g_c, that is $\{g | g > g_c\}$.

A practical calculation of g_c and the construction of the critical subnetwork proceeds as follows. Assume that the conductivity due to the fraction $\{g | g > g_c\}$ can be written as

$$\sigma(R) = \frac{R}{R_c} S(R) \phi \sigma_w$$

where $S(R)$ is the fraction of connected porosity that is accessible through pore throats of radius greater than or equal to R. This relation arises from the result that the conductance of a capillary of cross-section S and length l is $g = \dfrac{\sigma S}{l}$ and the approximation $\dfrac{S}{l} = R$. Note that because $S(R)$ decreases as R increases, there is a maximum in the product $RS(R)$ at $R = R_{max}$. In the first approximation the conductivity of the rock is taken as

$$\sigma = \sigma(R_{max}).$$

Katz and Thompson (1987) applied this approach to a suite of sandstones. The function $S(R)$ was determined by mercury porosimetry (equs. II.13 and II.7), from which R_{max}, the maximum of the function $RS(R)$, was determined. Their predicted conductivities compared very well with experimentally measured values.

One can show however that this method is accurate only in the case where the conductance distribution is very wide and decreases sufficiently rapidly (David et al., 1990). This last condition, a rapidly decreasing conductance distribution, is necessary for a dominant subnetwork to appear and is often found to be the case for fractured rocks. A computation only on the subnetwork simplifies the numerical computation of σ_{eff} for a network of conductances.

Relations between Conductivity-Permeability-Porosity

Research on the relation between permeability-conductivity and porosity-conductivity is motivated by the economic importance of the parameters k and ϕ. Because conductivity measurements in the laboratory or in situ are relatively easy, the ability to deduce k from σ constitutes a goal of major importance. The importance of this problem is precisely the reason why we shall examine a number of these approaches.

The calculations of the tube and fissure models presented in chapter V and the following chapter (Guéguen and Dienes, 1989) lead to the relation:

$$F = \frac{4}{f\phi} \qquad\qquad \text{VIII.22}$$

where VIII.20 is one approximation. Combining this relation and the results of table V.1 one obtains various relations between F, k, and ϕ. For example, the tube model implies:

$$kF = \frac{\overline{r^2}}{8}. \qquad\qquad \text{VIII.23}$$

The tube (or fissure) model can be interpreted in terms of three microvariables: $\overline{d}$ (length), $\overline{r}$ (radius), $\overline{l}$ (spacing). Because F, k, and ϕ all depend on these three microvariables, in theory it is possible to deduce $\overline{d}$, $\overline{r}$, and $\overline{l}$ from measurements of F, k, and ϕ. This is only true if the three quantities F, k, and ϕ are independent. Equation VIII.22

shows that this is not true, and thus it is not possible to obtain a unique relation $k = k(F, \phi)$, unless there exist other independent relations between the microvariables. Other similar relationships are possible which reflect the history of the rock. The parameters l, d, r can change with time.

Katz and Thompson (1987) derived a relation between permeability and conductivity using the approach of Ambegaokar which we discussed previously. Their method consists of writing two analogous relations for σ and k based on their scaling properties: the electrical conductance being proportional to $RS(R)$ and the permeability to $R^3S(R)$.

$$\frac{\sigma}{\sigma_w} = \phi\left(\frac{R_{max}^e}{R_c}\right) S(R_{max}^e)$$

$$k = \frac{1}{89} \phi(R_{max}^h)^2 \left(\frac{R_{max}^h}{R_c}\right) S(R_{max}^h).$$

Here R_{max}^e and R_{max}^h are the two radii corresponding to the maximums of the functions $RS(R)$ and $R^3S(R)$. Using results of percolation theory, Katz and Thompson related R_{max}^e and R_{max}^h to the critical radius R_c and arrived at a relation analogous to VIII.23.

$$kF = \frac{(R_c)^2}{226}.$$

In this manner two fundamental transport properties are related through a length parameter R_c, which is associated with a percolation problem. Note that R_c is the critical pore throat radius, while $\bar{r}$ is the mean radius.

It is equally possible to provide an interpretation of Archie's law VIII.16 from the preceding considerations. In utilizing VIII.22 for a tube model where the density of tubes (parameter $\bar{l}$) varies, one obtains

$$F \sim (\phi - \phi_c)^{-2} \phi^{-1} \qquad \text{with } \phi_c = \frac{2\pi}{3} \frac{\bar{r}}{\bar{d}}.$$

This result approaches Archie's law, suggesting that this empirical relation is due to a supplementary condition on the microvariables. During diagenesis $\bar{d}, \bar{r}$, and $\bar{l}$ all evolve. In the above example $\bar{r}$ and $\bar{d}$ are considered constant and $\bar{l} = \bar{l}(t)$ varies.

In order to simulate the effects of compaction and cementation in the course of time, many authors have sintered glass beads of identical or variable diameters. For these cases the pore geometry evolves. The

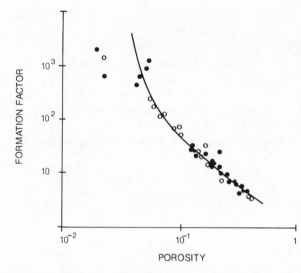

Fig. VIII.17. Formation factor-porosity relation obtained for sintered glass beads. Solid line is a theoretical model. (After Roberts, J. N., and L. M. Schwartz, 1985, *Phys. Rev. B.* 31:5990.) ○ Spheres of identical radii. • Spheres of different radii (bimodal distribution).

results of these experiments and numerical simulations have shown behavior approaching that of Archie's law (fig. VIII.17).

PROBLEMS

VIII.a. Water Saturation

(1) By utilizing Archie's two laws, show that it is possible to express S_w as a function of ρ_t (resistivity of the partially saturated rock) and ϕ: $S_w = f(\rho_t, \phi)$ where f is parameterized by m, n, and ρ_w.

(2) Extend this relation to the case where the rock contains a non-negligible fraction of clay minerals.

VIII.b. Tube and Fissure Model

(1) Utilizing the results of chapter V (table V.I) and equation VIII.21, show that it is possible to obtain Archie's law in the form $F = (\phi - \phi_c)^{-2}\phi^{-1}$ by assuming that for the series of rocks considered, the density of tubes and fissures varies. Calculate ϕ_c.

(2) Using the previous results, but now assuming that the tube radius r or fissure aperture w vary, show that one obtains a relation $k \sim F^{-n}$, with $n = 2$ (for tubes) and $n = 3$ (for fissures).

VIII.c. The Models of Kirkpatrick and Percolation

Assume that the medium can be described by a binary distribution of conductances:

$$g_{ij} = g_o \quad \text{with probability } p;$$

$$g_{ij} = 0 \quad \text{with probability } 1 - p.$$

Compute the effective conductance g_m and show that there exists a percolation threshold for g_m at the critical value $p_c = 2/z$, z being the number of nearest neighbors.

REFERENCES

Ambegaokar, V., B. I. Halperin, and J. S. Langer. 1971. Hopping Conduction in Disordered Systems. *Phys. Rev. B* 4:2612–20.

Charlaix, E., E. Guyon, and S. Roux. 1987. Permeability of a Random Array of Fractures of Widely Varying Apertures. *Transport in Porous Media*, 2:31–43.

David, C., Y. Guéguen, and G. Pampoukis. 1990. The Effective Medium Theory Applied to the Transport Properties of Rocks. *J. Geophys. Res.*, 95:6993–7006.

Guéguen, Y., and J. Dienes. 1989. Transport Properties of Rocks From Statistics and Percolation. *Math. Geol.* 21:1–13.

Katz, A. J., and A. H. Thompson. 1987. Prediction of Rock Electrical Conductivity From Mercury Injection Measurements. *J. Geophys. Res.* 92:599–607.

Kirkpatrick, S. 1971. Classical Transport in Disordered Media: Scaling and Effective Medium Theories. *Phys. Rev. Lett.* 27:1722.

———. 1973. Percolation and Conduction. *Rev. Mod. Phys* 45:574.

Lenormand, R. 1987. Statistical Physics and Immiscible Displacement Through Porous Media. *AIP Conference Proceedings* 154:98–115. Edited by J. R. Banavar, J. Koplik, and K. W. Winkler.

Waxman, M. H., and L.J.M. Smits. 1968. Electrical Conductivities in Oil Bearing Shaly-Sands. *Am. Inst. Min. Metall. and Petr. Eng.* 243:107–22, pt. II.

IX. Dielectric Properties

IN RECENT YEARS there has been a growing emphasis on electrical methods in geophysical exploration. Electromagnetic measurement techniques are now used routinely to probe the earth over 17 orders of magnitude in frequency, from 10^{-6} cycles per second (or Hertz) for deep magnetic sounding to microwave frequencies ($10^9 - 10^{11}$ Hz). At high frequencies microwave radiometry can view the Earth's layered structure over distances on the order of a fraction of a meter, while at the lowest frequencies deep magnetic sounding can penetrate more than a hundred kilometers. Low frequencies provide us with information on the conductivity structure (chap. VIII), while at very high frequencies the dielectric properties are measured. This wide spectrum of frequencies and wavelengths implies a great diversity of electrical sources and measurement devices that are used in electrical methods. The multitude of scales of applications and measurement techniques are much too vast to be included in a single chapter. But, irrespective of the application and the frequency of investigation, data interpretation in terms of rock properties requires a good understanding of how rocks and minerals respond to time-varying electromagnetic fields. Achieving a better understanding of the electromagnetic response of rocks in terms of the rock composition and microstructure is the aim of this chapter.

DIELECTRIC POLARIZATION

When an electric field is applied to a material, in addition to a current of free charges, there is a local redistribution of bound charges to new equilibrium positions. This phenomena of charge redistribution is called polarization and results from a number of distinct processes (fig. IX.1). An *electronic* contribution arises from the distortion of the electron shell relative to the atomic nucleus. *Ionic* contributions result from the relative displacement and deformation of charged ions with respect to each other. *Dipolar* polarization arises when molecules possessing a permanent electric moment (e.g., water molecules) are oriented along an applied field. In addition there can be a *space-charge* polarization due to local migration of charged particles. This latter polarization is a common phenomena in heterogenous materials such as porous rocks saturated with a brine.

ZERO FIELD   NON ZERO FIELD

◄— E

ELECTRONIC POLARIZATION

IONIC POLARIZATION

DIPOLAR POLARIZATION

SPACE CHARGES POLARIZATION

Fig. IX.1. The principal polarization processes. (After Von Hippel, A. R., 1954, *Dielectrics and Waves*. New York: John Wiley and Sons, pp. 21–25.)

Dielectric Permittivity

Separation of microscopic charges results in an induced field which tends to oppose the applied field. Classically this displacement is viewed as the motion of electrons, nuclei, and polar molecules from neutral equilibrium positions, to new equilibrium positions where the Coulomb forces between them balance the force of the applied field. For example, two electric charges of opposite polarity $\pm q$ when separated by a distance d, form an electric dipole of moment $q\vec{d}$. This dipole moment is a vector quantity of magnitude $q|\vec{d}|$ pointing from the negative to the positive charge. The polarization $\vec{P}$ of a material is defined as the induced electric dipole moment per unit volume. Macroscopically this effect is characterized through the coefficient ε, the permittivity of a medium, and the constitutive equation:

$$\vec{D} \equiv \varepsilon_o \vec{E} + \vec{P} \equiv \varepsilon \vec{E}. \qquad \text{IX.1}$$

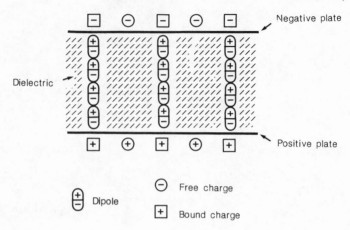

Fig. IX.2. Polarization on a dielectric placed between two parallel plates of a condenser.

Equation IX.1 relates the electric field $\vec{E}$ to the displacement field $\vec{D}$. ε_o is the permittivity in the absence of matter (in a vacuum): $\varepsilon_o = 8.854 \times 10^{-12}$ Farads/meter. The international system of MKS units will be used in this book. They are summarized below:

E Electric field intensity (volts/meters)
D Electric displacement (coulomb/m² = farad volt/m²)
P Polarization (coulomb/m² = farad volt/m²)
ε Dielectric permittivity (farads/meter)
C Capacitance (farad = sec²coul²/kg m²).

For a discussion of the units used in electromagnetic field problems see Von Hippel (1954).

Conceptually, measurements of polarization $\vec{P}$, displacement field $\vec{D}$, and dielectric constant ε can be made by measuring the charge on a condenser which is filled with the material of dielectric constant ε (Fig. IX.2). In the absence of a dielectric, a potential difference V between the two plates of the condenser produces a free charge density Q_f on the electrodes: $Q_f = CV$. When the dielectric is inserted between the plates, polarization charge densities $\pm P$ are induced on the two surfaces of the dielectric. These charge densities are neutralized by the flow of charge around the condenser circuit. The resultant charge density is $Q_f + Q_b$, where Q_b represents the bound charge necessary to neutralize the dielectric polarization. If A is the electrode area, the normal components of $\vec{D}$, $\vec{P}$ and $\varepsilon_o \vec{E}$ can be directly related to these charge densities.

$$\vec{D}\cdot\vec{n} \equiv \frac{Q_f + Q_b}{A}, \qquad \vec{P}\cdot\vec{n} \equiv \frac{Q_b}{A}, \qquad \varepsilon_o\vec{E}\cdot\vec{n} \equiv \frac{Q_f}{A}.$$

The dielectric constant κ is defined as the permittivity relative to the value in vacuum ε_o.

$$\kappa \equiv \frac{\varepsilon}{\varepsilon_o}. \qquad\qquad \text{IX.2}$$

κ is a dimensionless quantity equal to $\kappa = \dfrac{Q_f + Q_b}{Q_f}$. A linear approximation between $\vec{P}$ and $\vec{E}$ is applicable for most dielectrics:

$$\frac{\vec{P}}{\varepsilon_o} \equiv \chi\vec{E} \qquad \text{or} \qquad \varepsilon = \varepsilon_o(1 + \chi).$$

In general the parameters χ, κ and ε are tensors and can be complex quantities. The imaginary parts of these quantities represent energy dissipation during charge redistribution.

Frequency Dependence of the Permittivity

When the electric field (or potential across a condenser) varies sufficiently slowly, bound charges can keep up with the changing field and be in equilibrium with the instantaneous value of the field $\vec{E}$. The polarization will be in quasi-static equilibrium. But if the field $\vec{E}$ oscillates too rapidly, certain charge redistribution processes will not be able to keep up with the field and fully contribute to the polarization. The different processes considered previously are characterized by the amount of time (relaxation time) it takes for the polarization to be established. Because polarization phenomena are additive, the total polarization decreases as frequency increases. Thus measuring the frequency dependence of the permittivity is an important method for investigating and identifying the properties of rocks and minerals. By choosing different frequency ranges for the electromagnetic (EM) source, various components of a rock can be probed and identified.

Electrons having the smallest mass can respond to all time-varying fields up to and including optical frequencies. For all practical purposes *electron polarization* can be considered as constant and proportional to the number of electrons per unit volume. The electronic contribution to the permittivity is measured easily at optical frequencies, where it can be related to the index of refraction n:

$$\frac{\varepsilon}{\varepsilon_o} = n^2. \qquad\qquad \text{IX.3}$$

The relaxation time associated with electron displacements is on the order of 10^{-15} secs, which is considerably less than the period of radio waves. Electron polarization is independent of temperature except for the effect of thermal expansion, which decreases the number of electrons per unit volume (Polarization = dipole moment/unit volume). Pressure has the opposite effect of temperature, increasing the density and thus the permittivity. Electron polarization occurs in all materials, whether they are solid, liquid, or gas.

Ionic polarization is observed in crystalline as well as amorphous materials. It depends on atom type and their interactions with their neighbors. Because atomic masses are much larger than those of electrons, the relaxation times for ionic polarization are much longer than the relaxation times for electron polarization. This implies that ionic polarization is more important at lower frequencies. Even so, atomic motions normally occur in less than a microsecond for heavy atoms and approach periods associated with optical frequencies for light atoms (10^{-12} to 10^{-13} secs). At radio frequencies and below, both the atomic and electronic polarizations are normally present. The dielectric constants of most insulating minerals will only have atomic and electronic contributions and fall in the range of 4–15.

Homogeneous polar substances, such as water, contain molecules which possess permanent electric dipole moments. When an exterior electric field is applied, these polar molecules rotate attempting to align themselves along the direction of the field. Thermal agitation opposes this alignment and tends to maintain the molecules in a state of random orientation. As a consequence the degree of polarization in these types of materials depends strongly on temperature. Polar substances generally have large dielectric constants.

Inhomogeneous substances, particles of high conductivity imbedded in a continuous insulating matrix for example, can have very large polarizations due to *interface effects*. These effects depend strongly on microstructure and the constituent minerals. They are usually important only at much lower frequencies (1 kHz = 10^3 Hz). Low frequencies are required to allow time for migration of charges through the material. For example, when a field is applied to a porous medium containing an electrolytic solution, ions migrate in opposite directions and can accumulate at solid interfaces which block their motion. Such accumulations can result in large interfacial polarizations and dielectric constants. Processes of these types are grouped under the name Maxwell-Wagner effect. The classical Maxwell-Wagner response was developed to model a system in which a small volume fraction of one material was dispersed in a medium of different permittivity (or conductivity). Due to these processes, the dielectric constant of a water-rock system can approach

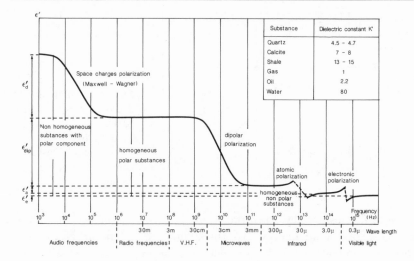

Fig. IX.3. Dielectric dispersion associated with different microscopic polarization processes.

values on the order of 1000, much larger than either that of water ($\kappa \sim 80$) or any mineral grain ($\kappa \sim 4{-}15$).

Figure IX.3 shows a generalized picture of the various polarization mechanisms and the frequency range over which they contribute. The dielectric constant decreases from its low-frequency value κ_o to a limiting value κ_∞ at high frequencies through several transitions. In the transition regions there is energy dissipation which is described through the complex dielectric constant κ: the real part κ' being the dielectric constant and the imaginary part κ'' describing energy dissipation. A classical model for analyzing the different transitions is due to Debye. This model, developed for dipolar polarization, has a single relaxation time τ which is the time necessary for establishing (or destroying) the polarization. The dielectric constant κ is a function of the angular frequency ω:

$$\kappa(\omega) = \kappa'(\omega) - i\kappa''(\omega) = \kappa_\infty + \frac{\kappa_o - \kappa_\infty}{1 + i\omega\tau}$$

where κ_o and κ_∞ are the limiting values of κ at low and high frequencies, respectively (when $\omega\tau \ll 1$ and $\omega\tau \gg 1$). The real and imaginary parts have the following simple analytical forms and are shown as solid lines in figure IX.4.

$$\kappa' = \kappa_\infty + \frac{\kappa_o - \kappa_\infty}{1 + (\omega\tau)^2}; \qquad \kappa'' = (\kappa_o - \kappa_\infty)\frac{\omega\tau}{1 + (\omega\tau)^2}. \qquad \text{IX.4}$$

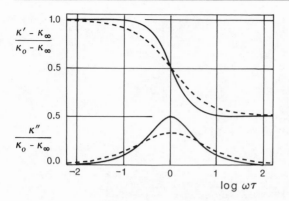

Fig. IX.4. The real (κ') and imaginary (κ'') parts of the dielectric constant. Solid curves— Debye model; Dashed curves—Cole-Cole model.

Note that essentially all of the decrease in the dielectric constant κ' and the maximum energy loss (maximum of κ'') occur around the critical frequency $\omega = 1/\tau$. This picture is formally analogous to Zener's model presented in chapter VII (table VII.6, fig. VII.9). In fact the analogy between these two cases is exact because they arise from the same model of relaxation. In this model the response of a medium to a small disturbance (mechanical constraint or an electric field) is assumed to be a linear function of the disturbance, which is composed of a part in phase (deformation proportional to the force or polarization proportional to the field) and a part out of phase by $\pi/2$ (derivative of the deformation proportional to the force or derivative of the polarization proportional to the field).

Experimental data indicate a broader frequency dependence than that predicted by the Debye model and suggest that more than one relaxation process is occurring. The dashed curves in figure IV.4 correspond to a model presented by Cole and Cole (1941) which simulates an ensemble of relaxation processes.

$$\kappa(\omega) = \kappa_\infty + \frac{\kappa_o - \kappa_\infty}{1 + (i\omega\tau)^{1-\alpha}}.$$

The parameter α varies from 0 to 1 with $\alpha = 0$ corresponding to the Debye model. The high-frequency dielectric constant κ_∞ is generally independent of temperature and the presence of water: κ_∞ depends only on the density and approaches the square of the index of refraction at optical frequencies ($\kappa_\infty = n^2$). The low frequency dielectric constant κ_o is larger due to polar components and generally has a strong temperature dependence.

Table IX.1. THE DIELECTRIC CONSTANT κ' FOR COMMON ROCK-FORMING
MINERALS AT HIGH FREQUENCIES

Mineral		Radio Frequencies	Optical Frequencies
Anhydrite, $CaSO_4$	Polycrys.	6.50	2.50
Gypsum, $CaSO_4 12H_2O$	*a* axis	11.2	2.31
	b axis	12.0	2.32
	c axis	5.4	2.34
Halite, NaCl		5.70–6.20	2.39
Aragonite, $CaCO_3$	*a* axis	6.46	2.34
	b axis	9.72	2.82
	c axis	7.55	2.84
Calcite, $CaCO_3$	Polycrys.	6.35	
	$\perp$ optic axis	—	2.21
	‖ optic axis	—	2.75
Dolomite, $CaMg(CO_3)$	$\perp$ optic axis	7.53	2.28
	‖ optic axis	6.11	2.85
Plagioclase Feldspar,			
$Ab_{98}An_2$-$Ab_{94}An_6$	Polycrys.	5.39–5.63	2.33–2.36
Ab_2An_{98}-Ab_7An_{93}	Polycrys.	7.05–7.14	2.49–2.51
Quartz, SiO_2	$\perp$ optic axis	4.96	2.36
	‖ optic axis	5.05	2.41

Note: For a more extensive compilation of dielectric constant data see the works of
Parkhomenko (1967), Olhoeft (1979, 1981), and Keller (1987).

Dielectric Constant of Minerals and Water

Values of the dielectric constant for common rock-forming minerals, at
optical and radio frequencies, are shown in table IX.1. At optical
frequencies only the electronic component contributes and $\kappa' \approx n^2$.
This contribution is practically the same for all minerals. At radio
frequencies, which includes the range of most interest in geophysical
exploration, atomic polarization is significant. Dielectric constants of
minerals range over two orders of magnitude. But, for most common
rock-forming minerals, this variation is less than a factor of 10. The
dielectric constant of most insulators is a function of the polarizability
of each particle and the number of particles per unit volume. This
suggests that there should exist a correlation between dielectric constant and bulk density. Such a correlation can be seen in the data shown
in figure IX.5, which includes data on minerals with different compositions (oxides, silicates, carbonates) and different types of bonding (ionic

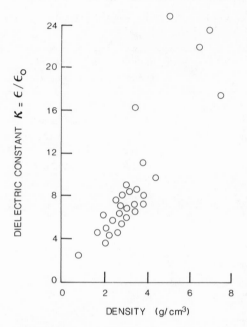

Fig. IX.5. Correlation between dielectric constant and density. (After Keller, G., 1987, Rock and Mineral Properties, in *Electromagnetic Methods in Applied Geophsics-Theory*, vol. 1, ed. M. N. Nabighian, Soc. of Expl. Geophys., Tulsa, Okla.

and covalent). Many oxides and sulfides have large dielectric constants due to the high polarizability of the oxygen and sulfur atoms. Semiconductors and oxides of metals (hematite, magnetite) are also characterized by particularly high values due to electron conduction.

Water is a very important component of crustal rocks. To interpret rock data it is essential to have a good understanding of the dielectric properties of water and its dependence on temperature, salinity, and frequency. Being composed of polar molecules, pure water illustrates the classical Debye frequency dependence shown in figure IX.6. Peak absorption of energy occurs at a frequency $\omega = 1/\tau \sim 10^{10}$ Hz at room temperature. Below this frequency κ' is independent of frequency but varies with temperature: $\kappa' = \kappa_o(T)$. When T increases, κ_o decreases (fig. IX.7). This result conforms with the fact that thermal agitation opposes the alignment of polar molecules along the field. As the temperature is lowered below zero °C, there is a large increase (by a factor of 10^7) in the dipolar relaxation time of ice. The relaxation time of ice increases as the temperature decreases and around -60 °C the contribution of dipolar relaxation to κ' is almost negligible (fig. IX.8).

The dielectric constant of water also depends on the concentration of impurities. Because water molecules have one of the highest polarizabilities of any liquid, addition of any "impurities" will reduce the average volume polarizability and thus the low frequency dielectric constant κ_o.

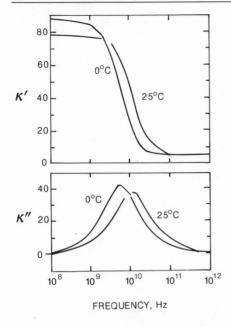

Fig. IX.6. Dielectric relaxation spectra for water at 0° and 25°C.

The high-frequency response, κ_∞, is not appreciably changed by salt impurities. The major effect of salt impurities is to significantly increase the water conductivity and thus energy dissipation at low frequencies.

Displacement Currents and Dielectric Loss

Maxwell's equations distinguish between electrical currents in the usual sense and displacement currents. Displacement currents do not exist in

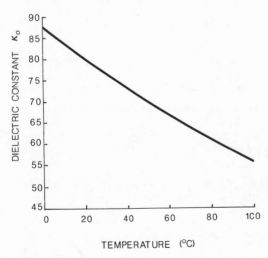

Fig. IX.7. Variation of the static dielectric constant of water κ_o as a function of temperature. (After Eisenberg, D., and Kauzman, W., 1969, *The Structure and Properties of Water*. Oxford University Press.)

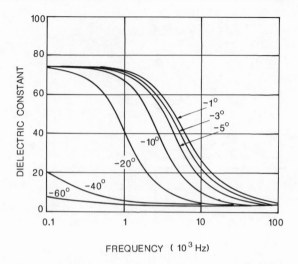

FREQUENCY (10^3 Hz)

Fig. IX.8. Dielectric constant of ice as function of temperature and frequency. (After Eisenberg, D., and Kauzman, W., 1969, *The Structure and Properties of Water*. Oxford University Press.)

a static electric field. They only exist when the field is varying in time. The total current $\vec{J_T}$ is thus a sum of two contributions:

$$\vec{J_T} = \sigma\vec{E} + \frac{\partial\vec{D}}{\partial t} = \sigma\vec{E} + \varepsilon\frac{\partial\vec{E}}{\partial t}. \qquad \text{IX.5}$$

The term $\sigma\vec{E}$ represents the electrical current in the usual sense, the conduction current discussed in chapter VIII. The term $\dfrac{\partial\vec{D}}{\partial t}$ is the displacement current, a part of which $\left(\text{a term } \dfrac{\partial\vec{P}}{\partial t}\right)$ is due to bound charges. For oscillatory electric fields of the form $E \sim e^{i\omega t}$, the total current can be written as

$$\vec{J_T} = (\sigma + i\omega\varepsilon)\vec{E} \equiv \sigma_T\vec{E} \equiv \varepsilon_T\frac{\partial\vec{E}}{\partial t}. \qquad \text{IX.6}$$

The total electric conductivity, σ_T, is a complex quantity and is related to the total dielectric permittivity, ε_T, through $\sigma_T = i\omega\varepsilon_T$. The permittivity ε is itself a complex quantity

$$\varepsilon \equiv \varepsilon' - i\varepsilon''$$

where the imaginary part ε'' is associated with energy dissipation. Therefore σ_T and ε_T can be separated into their real and imaginary parts in the following manner:

$$\sigma_T = (\sigma + \omega\varepsilon'') + i\omega\varepsilon'$$

$$\varepsilon_T = \varepsilon' - i\left(\frac{\omega\varepsilon'' + \sigma}{\omega}\right). \qquad \text{IX.7}$$

The constitutive equation for the total current can be written in two equivalent linear forms: either utilizing ε_T or σ_T to characterize the medium. The quantity $\omega\varepsilon''$ is equivalent to a conductivity. It contributes to energy dissipation by displacement currents just as σ contributes to energy dissipation by conduction currents. Energy dissipation is often characterized by the loss tangent: the ratio of the real part of σ_T to imaginary (or imaginary part of ε_T to real).

$$\tan \delta \equiv \frac{\sigma + \omega\varepsilon''}{\omega\varepsilon'} = \frac{\sigma}{\omega\kappa'\varepsilon_o} + \frac{\kappa''}{\kappa'}.$$

For a material of finite conductivity, dissipation of energy is very important at *low* frequencies ($\omega \ll \sigma/\kappa'\varepsilon_o$). In addition there can be dielectric losses $\dfrac{\kappa''}{\kappa'}$ due to polarization relaxation, which occur around the critical frequency τ^{-1} of the relaxation process. Figure IX.9 shows schematically the frequency dependence of the real and imaginary parts of ε_T.

The total dielectric response of a medium, which is defined by ε_T, is often interpreted in terms of an equivalent electrical circuit. Note that equation IX.6 can be viewed as the simple circuit shown in figure IX.10a. The circuit is composed of a capacitance C in parallel with a resistance R (conductance $G = 1/R$) and is driven by an oscillating voltage $V \sim e^{i\omega t}$. The total current I separates into a charging current I_c through the capacitor and a current I_G through the resistance.

$$I_c = C\frac{dV}{dt} = i\omega CV, \qquad I_G = GV$$

$$I = I_C + I_G = (i\omega C + G)V = \left(C - i\frac{G}{\omega}\right)\frac{dV}{dt}. \qquad \text{IX.8}$$

By drawing the analogy between variables $\{J_T, E, \sigma, \varepsilon\}$ of the real medium and those of the electrical circuit $\{I, V, G, C\}$, it can be seen that the set of equations IX.6 is formally equivalent to IX.8. Although several useful analogies can be drawn, one should not conclude that the behavior of a dielectric material is equivalent to this type of circuit. One

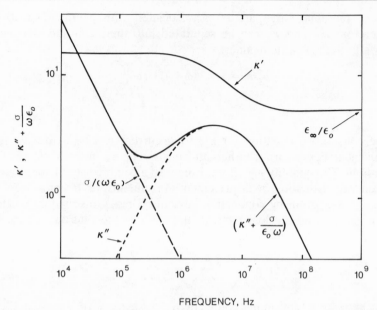

FREQUENCY, Hz

Fig. IX.9. General behavior of the real (κ') and imaginary $\left(\kappa'' + \dfrac{\sigma}{\omega \varepsilon_o}\right)$ parts of the total dielectric constant $\varepsilon_T/\varepsilon_o$. Curves are for a rock partially saturated with water. The relative importance of the different effects depends on the amount of saturation (higher saturations significantly increase σ) and the conductivity of the solution.

major difference is that G and C are simple circuit elements (constants) while σ and ε are complex quantities, which vary with frequency. The complex impedance of the circuit IX.10a is given by

$$\frac{1}{Z} = \frac{1}{Z_1} + \frac{1}{Z_2}; \qquad Z_1 = R, \qquad \frac{1}{Z_2} = i\omega C.$$

Thus

$$Z = Z' - iZ'' \qquad \text{with } Z' = \frac{R}{1 + (RC\omega)^2} \quad \text{and} \quad Z'' = \frac{R^2 C \omega}{1 + (RC\omega)^2}.$$

The resonant frequency of this circuit is given by $\omega\tau = 1$ with $\tau = RC$. It is common to exhibit the values of Z', Z'' in the complex plane with ω as the parameter. Using the preceding relations, one obtains a circle of radius R centered on the Z' axis. A maximum value for Z'' is attained for $\omega\tau = 1$ (fig. IX.10b). If the data is presented in the complex admittance plane $A = Z^{-1}$, the circle is deformed into a line (fig. IX.10c). This type of representation, or Cole-Cole diagram, permits a

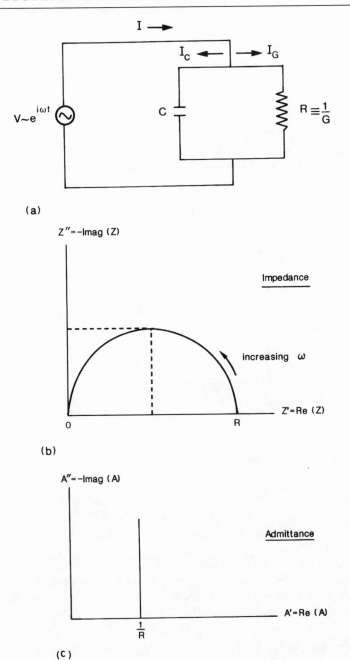

(a)

(b)

(c)

Fig. IX.10. (*a*) Equivalent circuit for an insulating dielectric. (*b*) Cole-Cole diagram for a circuit; using complex impedances. (*c*) Cole-Cole diagram for a circuit; using complex admittances.

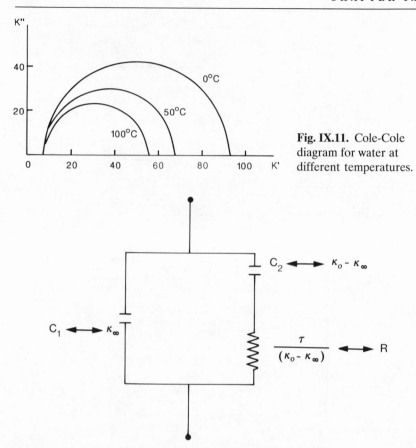

Fig. IX.11. Cole-Cole diagram for water at different temperatures.

Fig. IX.12. Equivalent electrical circuit for the Debye model.

clear visualization of the components around the critical frequency. The formal equivalence between equations IX.7 and IX.8 allows analogue Cole-Cole diagrams for both ε_T and σ_T. Figure IX.11 shows such a diagram for water.

The analogy between circuits and dielectric can be extended. For example, the circuit presented in figure IX.12 has the analogue components of the Debye model (equ. IX.4) and is the electrical equivalent of Zener's model shown in figure VII.9.

ELECTROMAGNETIC WAVES

In this section we will identify the relationships between velocity and attenuation of electromagnetic waves and the dielectric properties of the medium through which they are propagating. When a time-varying electromagnetic field is applied, it induces currents which interact with the applied field modifying its strength and velocity. These modifications depend on the properties of the material $\{\sigma, \varepsilon, \mu\}$, where μ is the magnetic permeability.

We have established that the total current is the sum of the conduction and displacement currents (equ. IX.5). This leads to Maxwell's first equation:

$$\nabla \times \vec{H} = \vec{J}_T = \sigma \vec{E} + \varepsilon \frac{\partial \vec{E}}{\partial t}. \qquad \text{IX.9}$$

Equation IX.9 is a generalization of Ampere's Law for the case of time-varying electric fields. Maxwell's second equation is a generalization of Faraday's law of induction:

$$\nabla \times \vec{E} = -\frac{\partial \vec{B}}{\partial t} = -\mu_o \frac{\partial \vec{H}}{\partial t} \qquad \text{IX.10}$$

where μ_o is the magnetic permeability of a vacuum, $\vec{H}$ and $\vec{B}$ are the magnetic field and magnetic flux density (magnetic induction) (cf. chap. XI). Equations IX.9 and IX.10, in conjunction with

$$\nabla \cdot \vec{D} = \rho$$

$$\nabla \cdot \vec{B} = 0 \qquad \text{IX.11}$$

where ρ is the density of free charges, lead to the equation for the propagation of an electromagnetic field:

$$\nabla^2 \vec{E} = \sigma \mu_o \frac{\partial \vec{E}}{\partial t} + \varepsilon \mu_o \frac{\partial^2 \vec{E}}{\partial t^2}. \qquad \text{IX.12}$$

One obtains this equation by taking the curl of equations IX.9 and eliminating $(\nabla \times \vec{H})$ terms between IX.9 and IX.10. An identical equation can be obtained for $\vec{B}$. Solutions of IX.12, for imposed boundary conditions, constitute an important basis for interpreting electromagnetic data.

Propagation of Electromagnetic Waves

Solutions to Maxwell's equations in three dimensions are mathematically very complex. Thus to illustrate the important features we shall limit ourselves to the special case of a plane wave of angular frequency ω propagating in the x-direction. We shall assume that the electric field is polarized in the y-direction and the magnetic field in the z-direction:

$$E_y = E_o e^{i(\omega t - kx)}, \qquad H_z = \frac{E_o k}{\mu_o \omega} e^{i(\omega t - kx)}.$$

k is the complex wavenumber (often called the propagation constant). The wavenumber k is determined by the condition that the electric and magnetic fields satisfy Maxwell's equations. Substituting E_y and H_z into equation IX.12 one obtains:

$$k^2 = \left(\varepsilon - i\frac{\sigma}{\omega}\right)\mu_o\omega^2 = (1 - i\tan\delta)\mu_o\varepsilon'\omega^2 \qquad \text{IX.13}$$

where $\varepsilon = \varepsilon' - i\varepsilon''$. This relation between the wavenumber, angular frequency, and material properties contains the essential information concerning electromagnetic wave propagation in a medium.

A result of IX.13 is that the wavenumber k is in general a complex quantity:

$$k \equiv \alpha - i\beta \Rightarrow E_y = E_o e^{-\beta x}e^{i(\omega t - \alpha x)}.$$

The real part of k describes the propagation of the electric field and its oscillations through the factor $e^{i(\omega t - \alpha x)}$: $\alpha = 2\pi/\lambda$ where λ is the wavelength. The imaginary part of k describes how the amplitude of the field decreases with distance, $E_o e^{-\beta x}$. A reduction in amplitude is associated with energy dissipation: the conversion of electromagnetic energy into heat. Attenuation is often expressed in terms of the "induction skin depth": the distance over which the field strength is attenuated to $1/e$ of its original value. The induction skin depth is equal to $1/\beta$. Explicit expressions for the real and imaginary parts of the wavenumber are

$$\alpha = \omega\sqrt{\mu_o\varepsilon'}\sqrt{\frac{1 + \cos\delta}{2\cos\delta}} \qquad \beta = \omega\sqrt{\mu_o\varepsilon'}\sqrt{\frac{1 - \cos\delta}{2\cos\delta}}.$$

Attenuation of a wave between x_1 and x_2 is defined as the logarithm of the ratio of the energy flux at x_2 and x_1. Because the energy of a wave is proportional to the square of its amplitude, attenuation will be equal to the logarithm of the square of the amplitude ratio. The quantities α and β can be obtained by measuring the amplitude and phase at two locations.

$$\text{Amplitude Ratio} = e^{-\beta(x_2-x_1)} \qquad \text{Phase shift} = \alpha(x_2 - x_1).$$

From these two measurements, the values of (α, β) and $(\varepsilon', \varepsilon'')$ can be determined.

$$\varepsilon' = \frac{\alpha^2 - \beta^2}{\mu_o\omega^2}, \qquad \varepsilon'' + \frac{\sigma}{\omega} = \frac{2\alpha\beta}{\mu_o\omega^2}.$$

Energy dissipation is normally expressed by the loss tangent:

$$\tan \delta \equiv \frac{\sigma}{\omega \varepsilon'} + \frac{\varepsilon''}{\varepsilon'} = \frac{2\alpha\beta}{\alpha^2 - \beta^2}$$

while the propagation of the wave is characterized by the phase velocity v:

$$v = \frac{\omega}{\alpha} = \frac{1}{\sqrt{\mu_o \varepsilon'}} \sqrt{\frac{2\cos\delta}{1 + \cos\delta}} \;.$$

A medium is called dispersive if the phase velocity varies with frequency. Because in general ε' and ε'' vary with frequency, v also varies with ω. In the absence of all dissipation, $\delta = 0$,

$$v = \frac{1}{\sqrt{\mu_o \varepsilon_o \kappa'}} = \frac{c}{\sqrt{\kappa'}}$$

where c is the speed of propagation in a vacuum.

The average behavior of the propagation parameters as a function of frequency is illustrated by the results in figure IX.13. The phase velocity data is for two moist crystalline rocks and one partially moist soil sample (mainly sand). For frequencies above 10^6 Hz, all velocities are essentially constant but lower than c by a factor $1/\sqrt{\kappa'}$. At frequencies lower than 10^6 Hz, v decreases rapidly due to increasing dissipation. The parameter $1/\beta$ (induction skin depth) measures the depth of penetration of an electromagnetic wave into a medium. It determines the effective distance over which the medium can interact with the incident field and thus is an important parameter in geophysical exploration methods. For distances greater than $1/\beta$, attenuation is very large and the amplitude is almost negligible. Figure IX.14 shows the penetration depth for these samples as a function of frequency.

High- and Low- Frequency Regimes

At very high frequencies ($\sigma \ll \omega \varepsilon'$), the conductive currents are negligible as compared to displacement currents. Because ε' (real part of ε) is normally much larger than ε'' (the imaginary part), the propagation of electromagnetic waves can occur without significant attenuation or dispersion. In this frequency regime, $\tan \delta = \varepsilon''/\varepsilon' \ll 1$, and

$$k = \omega(\mu_o \varepsilon')^{1/2}\left(1 - i\frac{\varepsilon''}{\varepsilon'}\right)^{1/2} \approx \omega(\mu_o \varepsilon')^{1/2}\left(1 - \frac{i}{2}\frac{\varepsilon''}{\varepsilon'}\right).$$

Thus measurement of the phase velocity is a direct measurement of the dielectric constant of the medium. The EM field satisfies the wave

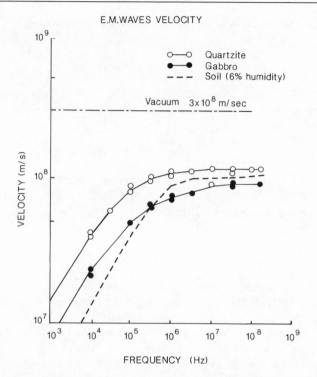

Fig. IX.13. Velocity of electromagnetic waves as a function of frequency for 3 different media (Katsube, T. J., and L. S. Collett, 1974, Electromagnetic Propagation Characteristics of Rocks, in *The Physics and Chemistry of Minerals and Rocks*, ed. R.G.J. Sterns, Wiley.)

equation, which in one dimension is

$$\frac{\partial^2 E_y}{\partial x^2} = \varepsilon\mu_o \frac{\partial^2 E_y}{\partial t^2} .$$

This high-frequency regime dominated by displacement currents is known as the *propagation regime* (fig. IX.15). EM field measurements can be utilized for determining distances to rock masses which reflect the waves due to contrasts in dielectric constant (and in phase velocity). There is a complete analogy with seismic methods. An example is radar sounding which is used for localizing cavities or fractures in rock masses.

At low frequencies ($\sigma \gg \omega\varepsilon'$), the displacement currents are negligible in comparison to conductive currents. For earth materials this

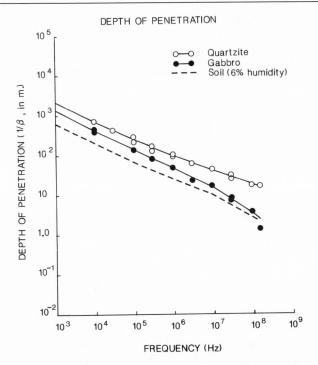

DEPTH OF PENETRATION

○—○ Quartzite
●—● Gabbro
– – – Soil (6% humidity)

DEPTH OF PENETRATION (1/β, in m.)

FREQUENCY (Hz)

Fig. IX.14. Depth of penetration (induction skin depth) $1/\beta$ as a function of frequency for 3 different media. (Katsube, T. J., and L. S. Collett, 1974, Electromagnetic Propagation Characteristics of Rocks, in *The Physics and Chemistry of Minerals and Rocks*, ed. R.G.J. Sterns, Wiley.)

condition ($\sigma \gg \omega \varepsilon'$) occurs at frequencies less than 10^5 Hz (cf. fig. IX.15). In this case the wavenumber can be expressed as:

$$k = (1 - i)\left(\frac{\omega \mu_o \sigma}{2}\right)^{1/2}.$$

At these frequencies the wavelength and penetration depth are of the same magnitude, and the amplitude decreases by a factor of $e^{-2\pi}$ over a distance of one wavelength. This implies that oscillatory EM fields can only penetrate a distance of a few wavelengths (or a few skin depths). To achieve equal penetration in very conductive media, it is necessary to use very low frequencies f, $\beta^{-1} \sim (f\sigma)^{-1/2}$. To make measurements at these low frequencies, many applications use impulsive field sources rather than oscillatory sources. If a one-dimensional EM pulse is emitted from a source at $z = 0$ at $t = 0$, the measured field a distance z from the source varies with time as shown in figure IX.16a. Note that

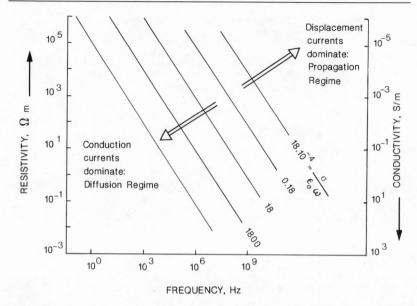

Fig. IX.15. Illustration of the high and low frequency regimes in the (resistivity, frequency) plane. The plane is separated into (1) the propagation regime at high frequencies, and (2) the diffusion regime at low frequencies.

the field is "spread" and arrives over a long time interval. The maximum is attained after a time Δt

$$\Delta t = \frac{\mu_o \sigma z^2}{6}.$$

If on the other hand the field is measured at a fixed time t as a function of distance z, it will appear as in figure IX.16b. The location z_m where the field reaches a maximum is given by

$$z_m = \left(\frac{2t}{\mu_o \sigma} \right)^{1/2}.$$

One observes a $t^{1/2}$ dependence, which is characteristic of a diffusive process. This is the *diffusive regime* and the EM field satisfies the diffusion equation. In one dimension:

$$\frac{\partial^2 E_y}{\partial x^2} = \sigma \mu_o \frac{\partial E_y}{\partial t}.$$

The maximum of the pulse travels with a velocity $\dfrac{dz_m}{dt} = (2\mu_o \sigma t)^{-1/2}$ and its amplitude varies as $1/t$. Electromagnetic measurements in the

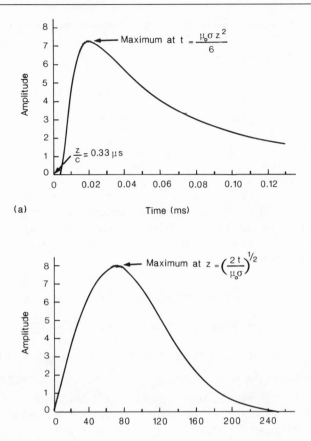

Fig. IX.16. "Diffusion" of an electromagnetic pulse emitted at $x = 0$ at $t = 0$. (a) Amplitude as a function of time at $z = 100$ m. (b) Amplitude as a function of distance at $t = 0.03$ ms. The medium has conductivity $\sigma = 0.01$ S/m. In a vacuum the field would be displaced as a pulse with a velocity of 300 m/μs.

diffusive (or inductive) regime provide information about the medium conductivity. Electromagnetic methods utilized are magnetotellurics, planetary sounding, etc.

DIELECTRIC PROPERTIES OF ROCKS

We have seen that each component of an aggregate can itself be a complex material characterized by macroscopic electrical parameters $\{\sigma, \varepsilon, \mu\}$. The focus here will be how the global rock properties depend on those of its constituents. In the static regime, the dielectric properties of a rock are obtained from the microscopic relation for the displacement field (in the absence of free charges).

$$\vec{\nabla} \cdot \vec{D} = \vec{\nabla} \cdot (\varepsilon \vec{E}) = -\vec{\nabla} \cdot (\varepsilon \vec{\nabla} V) = 0.$$

This equation is formally identical to that for the conductivity:

$$\vec{\nabla} \cdot \vec{J} = \vec{\nabla} \cdot (\sigma \vec{E}) = 0.$$

This implies that the computation of effective permittivity is mathematically identical to that of effective conductivity (cf. chap. VIII). All of the concepts introduced for the effective conductivity apply here for the effective permittivity. The only difference is that the contrast in permittivity between materials is not as great as that for conductivity. The effective permittivity ε_{eff} of a medium is defined by:

$$\langle \vec{D} \rangle = \langle \varepsilon \vec{E} \rangle \equiv \varepsilon_{\text{eff}} \langle \vec{E} \rangle. \qquad \text{IX.14}$$

Brackets denote an average over a representative volume (defined in chap. III).

How is this result changed when we have time-varying electric fields? First, a complex dielectric constant $\varepsilon_T = \left(\varepsilon - i\dfrac{\sigma}{\omega} \right)$ is required and not simply ε. In addition the static equations just discussed are no longer applicable. The electric and magnetic fields are coupled and must satisfy Maxwell's equations. Because ε_T varies from grain to grain, so do the time-varying fields $\vec{E}$ and $\vec{H}$. These small-scale fluctuations can be eliminated by an appropriate volume averaging to reveal the systematic variations on a scale much larger than the grain size. As previously discussed, averaging is over a representative volume V, such that:

$$\text{Grain size } (1\text{--}10^3 \mu m) \ll V \ll \text{Wavelength.}$$

This condition is almost always satisfied for frequencies less than optical frequencies. Maxwell's equations will then have the same mathematical form as on the microscale, but the propagation of an average electric and magnetic field will be described by an effective frequency-dependent permittivity ε_T^*. The effective permittivity ε_T^* is defined by the average:

$$\langle \vec{J}_T \rangle = \left\langle \left(\varepsilon - i\frac{\sigma}{\omega} \right) \frac{\partial \vec{E}}{\partial t} \right\rangle = \left\langle \varepsilon_T \frac{\partial \vec{E}}{\partial t} \right\rangle \equiv \varepsilon_T^* \left\langle \frac{\partial \vec{E}}{\partial t} \right\rangle.$$

We will use the relative dielectric "constant" κ^*:

$$\kappa^* \equiv \frac{\varepsilon_T^*}{\varepsilon_o} = \kappa' - i\left(\kappa'' + \frac{\sigma}{\omega\varepsilon_o}\right).$$

κ' is the effective dielectric constant of the rock, κ'' describes the effective dissipation due to displacement currents, and σ is the effective conductivity. In general all three response functions $\{\kappa', \kappa'', \sigma\}$ vary with frequency and depend on rock composition and microstructure. All three can be obtained through various theoretical methods. But these theoretical methods have a severe constraint: they must be capable of predicting the correct dielectric constant, conductivity, and energy dissipation over an extremely wide frequency range!

Model for a Water-Saturated Rock

We have seen in chapter VIII that the electrical conductivity of porous rocks is totally controlled by its water content. Although the contrast in permittivity is not as great as for conductivity, there is still a significant contrast at lower frequencies between most rock-forming minerals ($\kappa' \approx 5$) and water ($\kappa' \approx 80$). This contrast is due to the dipolar polarization in water and implies that the frequency dependence between water-saturated and dry rocks will be significantly different. Table IX.2 lists values of the low-frequency dielectric constant of several common rock-forming minerals and saturating fluids.

Because the dielectric constants of gas and oil are much lower than that of water, dielectric measurements can be used to determine relative saturations of oil and water. Note also that at low frequencies, the dielectric constant of a water-saturated rock is often much larger than any one of its constituents. Values as high as 1000 have been measured! Such high values are indicative of space-charge polarization due to microstructure.

Table IX.2. THE LOW-FREQUENCY DIELECTRIC CONSTANT OF COMMON ROCK-FORMING MINERALS AND SATURATING FLUIDS

Substance	Dielectric Constant
Quartz	4.5–4.7
Calcite	7–8
Shale	13–15
Water	80
Oil	2.2
Gas	1

The dielectric response of a water-saturated rock can be accurately modeled as a two-component mixture: rock matrix (described by a dielectric constant κ_s) and conducting brine (described by a complex dielectric function $\kappa_w(\omega)$). At frequencies *below* the dipolar relaxation frequency ($\kappa_W'' \approx 0$), the real part of κ_w (that is, κ_w') is approximately constant, and κ_w will have the form

$$\kappa_w = \kappa_w'(1 - i \tan \delta_w) \cong \kappa_w' - i\frac{\sigma_w}{\varepsilon_o \omega}.$$

(Note that κ_w now includes the term $\sigma_w/\omega\varepsilon_o$ in its definition.) At these frequencies dissipation is primarily due to water conductivity ($\sigma_w/\omega\varepsilon_o \gg \kappa_w'', \kappa_s''$). The properties of the mineral matrix are adequately described by a frequency independent, real-valued dielectric constant $\kappa_s = \kappa_s'$. Thus the rock can be characterized by three frequency independent, real parameters κ_w', κ_s', and σ_w. At low frequencies, the effective dielectric constant of a rock takes the form

$$\kappa^* \equiv \kappa' - i\frac{\sigma}{\varepsilon_o \omega}.$$

Here σ is the effective conductivity of the medium and the real part κ' is the effective dielectric constant of the composite medium. The two parameters, κ' and σ, depend on κ_w', κ_s', σ_w, and ω.

A useful scaling property of the effective dielectric constant κ^* is that if both κ_w and κ_s are multiplied by a constant λ, then so is κ^*.

$$\kappa^*(\lambda\kappa_w, \lambda\kappa_s) = \lambda\kappa^*(\kappa_w, \kappa_s) = \lambda\kappa_s\kappa^*(\kappa_w/\kappa_s).$$

This implies that the ratio κ^*/κ_s only depends on the ratio κ_w/κ_s and the volume fractions of the two phases:

$$\frac{\kappa^*}{\kappa_s} = F(\kappa_w/\kappa_s, \phi)$$

with

$$\frac{\kappa_w}{\kappa_s} = \frac{\kappa_w'}{\kappa_s'}(1 - i \tan \delta_w); \qquad \tan \delta_w = \frac{\sigma_w}{\omega\kappa_w'\varepsilon_o}.$$

The functional form of F is determined by rock microstructure.

An often-used example of F is

$$\sqrt{\kappa^*} = \phi\sqrt{\kappa_w} + (1 - \phi)\sqrt{\kappa_s}.$$

This empirical formula is found to be reasonably accurate at high frequencies (near 1000 MHz) and is called the "Complex refractive index method" (CRIM). At high frequencies the square root of the dielectric constant is inversely proportional to the phase velocity, $v = c/\sqrt{\kappa^*}$. Thus the preceding equation represents a time average equation (analogous to the Wyllie equation in chap. VII): the travel time of an electromagnetic wave, through a composite of water and rock of length L, is equal to the sum of travel times through the fluid {length ϕL} and solid {length $(1 - \phi)L$}.

$$\frac{L}{v} = \frac{L\sqrt{\kappa^*}}{c} = \frac{L\phi\sqrt{\kappa_w}}{c} + \frac{L(1 - \phi)\sqrt{\kappa_s}}{c}.$$

The scaling behavior of $F(\kappa_w/\kappa_s) = \kappa^*/\kappa_s$ is more apparent when the preceding equation is rearranged into the equivalent form:

$$\frac{\kappa^*}{\kappa_s} = \phi^2 \frac{\kappa_w}{\kappa_s} + (1 - \phi)^2 + 2\phi(1 - \phi)\sqrt{\frac{\kappa_w}{\kappa_s}}.$$

Separating κ^* into its real and imaginary parts, the effective dielectric constant κ' and effective conductivity σ can be obtained:

$$\kappa' = \phi^2 \kappa_w' + (1 - \phi)^2 \kappa_s' + 2\phi(1 - \phi)\sqrt{\kappa_w' \kappa_s'} \frac{\cos(\delta_w/2)}{\sqrt{\cos \delta_w}}$$

$$\frac{\sigma}{\sigma_w} = \phi^2 + 2\phi(1 - \phi)\sqrt{\frac{\kappa_s'}{\kappa_w'}} \frac{\sqrt{\cos \delta_w}}{2\cos(\delta_w/2)}.$$

At frequencies below the dipolar relaxation frequency of water, the loss angle δ_w, is the only frequency-dependent parameter. In the limit $\omega \to 0$, $\delta_w \to \pi/2$, and the conductivity approaches the result $\sigma = \sigma_w \phi^2$ (Archie's Law with exponent 2). At higher frequencies, σ increases with ω.

The preceding equation predicts very large values for κ' at low frequencies. Because $\kappa' \sim 1/\sqrt{\cos \delta_w}$, it can become much larger than the dielectric constant of water or the solid phase! This dielectric enhancement is a new feature that was not possible for the other effective transport properties discussed previously. In the high frequency limit ($\omega \to \infty$ and $\delta_w \to 0$) conductive dissipation becomes negligible and κ^* becomes a real number,

$$\frac{\kappa_w}{\kappa_s} \to \frac{\kappa_w'}{\kappa_s'}.$$

A second example of F which is currently utilized is given by the Lichtneker-Rother formula:

$$\kappa^* = \left(x_1(\kappa_1)^s + x_2(\kappa_2)^s + \cdots + x_n(\kappa_n)^s\right)^{1/s}$$
$$-1 \leq s \leq 1.$$

Again, κ_i and x_i represent the dielectric constant and volume fraction of the i-th component. The value $s = 1$, correspond to a model of dielectrics in parallel, while a value $s = -1$ corresponds to dielectrics arranged in series. A value of $s = 1/2$ leads to the time average equation discussed previously. Intermediate values of s are meant to simulate geometries intermediate to these two extremes (parallel/ series). Measurement of the dielectric constant κ' at a single frequency ω_o, determines the exponent s. It is of course not clear if this value of s determined at ω_o will fit the data over the full frequency range ($\omega \gg \omega_o$ and $\omega \ll \omega_o$). The Lichtneker-Rother formula is empirical. As with most empirical formulas, the relationship between physical property and rock microstructure is not clearly defined.

Dielectric Constants of Dry and Water-Saturated Rocks

The preceding analysis has shown the importance of frequency on dielectric properties. It is of particular importance to understand the frequency domain over which the microstructure most affects the dielectric properties.

Dielectric Constant of Dry Rocks

Normally only the atomic and electronic components of polarization are available for dry rocks. Thus there is minimal or no dispersion for frequencies lower than 10^{12} Hz, and the dielectric constant is usually independent of all parameters except bulk density. The correlation between dielectric constant and density observed for individual minerals carries over to dry rocks. Dielectric constant data of Olhoeft (1979) for 114 lunar samples, 261 monomineralic specimens, and 367 rocks shows that this data has a simple representation:

$$\kappa^* = \kappa' = (\kappa'_s)^{\rho/\rho_s} \approx (1.91)^{\rho}.$$

Here κ' and ρ represent the measured (effective) dielectric constant and density of the samples, and κ'_s and ρ_s the dielectric constant and density of the constituent minerals. The data which was not well represented by this empirical relation was for samples that contained a high percentage of water (as in montmorillonite clay for example) or conducting (magnetite) or semiconducting (pyrite and other sulfides) minerals. The empirical law $\kappa^* \approx (1.91)^{\rho}$ can arise from the equation

$$\kappa' = (\kappa'^{x_1}_1) \times (\kappa'^{x_2}_2) \times \cdots \times (\kappa'^{x_n}_n)$$

for a medium of n components with dielectric constants $\kappa'_1 \ldots \kappa'_n$ and volume fractions $x_1 \ldots x_s$. In the present case $n = 2$, the main constituent being the solid phase $\{\kappa'_s$ and $x_s\}$ and the second constituent being air $\{1, 1 - x_s\}$. The previous equation then simplifies to

$$\kappa' = (\kappa'_s)^{x_s} = (\kappa'_s)^{\rho/\rho_s}$$

where ρ and ρ_s are the average densities of the porous dry rock and mineral grains, respectively. The data was then used to determine the best value for $(\kappa'_s)^{1/\rho_s}$, which turned out to be 1.91.

Dielectric Constant of Water-Saturated Rocks

The presence of water introduces a frequency dependence into the dielectric reponse of a rock. With the addition of a few weight percent of water, the dielectric constant is noticeably increased at low frequencies. This observation is consistent with the process of dipolar polarization. In addition the electrical conductivity and energy dissipation increase by several orders of magnitude when water forms a percolating cluster through the rock. This increased dissipation by conduction makes it difficult in practice to measure the dielectric constant.

Figures IX.17a, b show the dielectric constant and loss tangent of an intact and powdered granite at room temperature. Both samples were measured in a vacuum and in an atmosphere of 30% humidity. The measurements in vacuum show minimal dispersion. Differences in dielectric constants between solid and powder can be directly attributed to differences in sample density (porosity). According to the relation $\kappa^* \approx (\kappa'_s)^{\rho/\rho_s}$, the loss angle δ of the solid sample is reduced to $\delta(\rho/\rho_s)$ by the frequency-independent factor (ρ/ρ_s) when the granite (density ρ_s) is transformed into a powder (density ρ). Vacuum data shown in figure IX.17 are in accord with this model. Note that in a vacuum, dielectric losses are very small for granite in both solid and powdered forms $(\kappa''/\kappa' \sim 10^{-2})$.

When the samples are exposed to air (30% humidity), there is a noticeable change in the electrical properties. In addition to the expected rise in conductivity, there is an increase in the dielectric constant at low frequencies due to the polarization of water molecules. Perhaps the most distinctive effect of adding water is that the solid and powder no longer have a similar frequency response, quite in contrast to the preceding case. Because the powder has a higher porosity and surface area, these differences may be attributed to the amount and state of adsorbed water on the mineral grains.

What frequency response can we expect when a rock, such as sandstone, is saturated with water? Because mineral grains exhibit minimal intrinsic dispersion, we expect the primary dispersion to occur around $\omega \sim 10^{10}$ Hz as predicted for pure water (fig. IX.6). At low frequencies, the dielectric constant should be larger in proportion to the volume

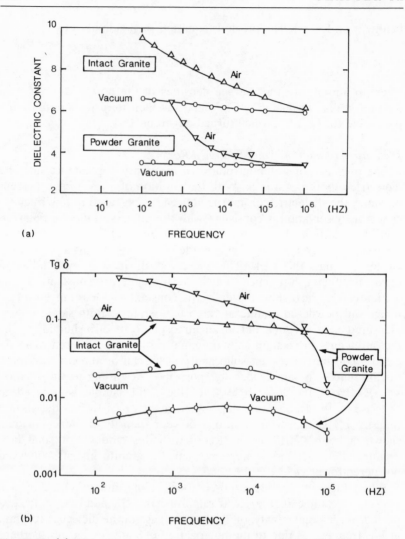

Fig. IX.17. (*a*) Dielectric constant of solid and powder granite, in a vacuum and in the presence of air (30% relative humidity) at ambient temperature. (*b*) Dielectric loss (tan δ) for the same granite and conditions. (After Strangway, D. W., W. B. Chapman, G. R. Olhoeft, and J. Carnes, 1972, *Earth Planet. Sci. Letters* 16:275.)

fraction of water that is present. We can also expect increased dissipation due to the ionic conductivity of saline water. Surface phenomena can noticeably modify these predictions. Figure IX.18 shows how the dielectric constant κ' varies for several sandstones of different porosities, saturated with fresh water. The constant κ' is greatly enhanced at

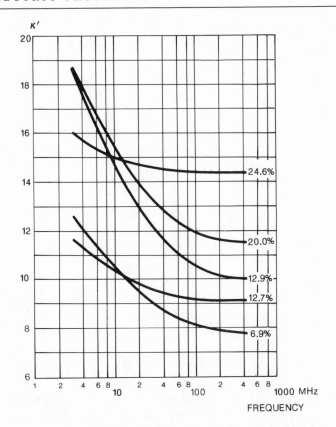

Fig. IX.18. Real part of the dielectric constant for fresh water saturated sandstones as a function of frequency. (After Poley, J. P., J. J. Nooteboom, and P. J. deWaal, 1978, *The Log Analyst* [May–June], p. 8.)

frequencies below 200 MHz. It is very clear that this low-frequency dispersion cannot be explained in terms of the properties of the individual components.

A clearer picture of this data can be attained if we view the components κ' and $\kappa'' + \sigma/\omega\varepsilon_o$ on an expanded frequency range. Figure IX.19 shows data on a saturated sandstone (porosity $\phi = 0.145$, salinity $= 4.3\%$ wt) over a frequency range from 1.5 kHz to 2400 MHz. Note that κ' reaches a value of 10^4 at ~ 2 kHz, a value much larger than either component. Because dipolar relaxation occurs around $\sim 10^{10}$ Hz, the low-frequency losses are due to brine conduction, as can be seen by the inverse frequency dependence $\sigma/\omega\varepsilon_o$ at low frequencies. Salinity has the effect of increasing conductivity and translating the linear portion to higher frequencies (fig. IX.19b). At frequencies above

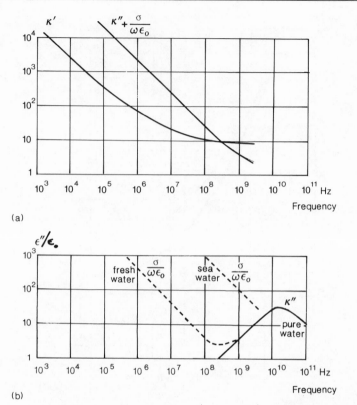

Fig. IX.19. (a) The real (κ') and imaginary $\left(\kappa'' + \dfrac{\sigma}{\omega\varepsilon_o} \right)$ parts of the dielectric constant as a function of frequency for sandstone saturated with saline water. (b) Dielectric κ'' and conduction $\dfrac{\sigma}{\omega\varepsilon_o}$ losses for fresh, pure, seawater. (After Poley, J. P., J. J. Nooteboom, and P. J. deWaal, 1978, *The Log Analyst* [May–June], p. 8.)

10^{10} Hz, conduction losses become small and κ' reaches a limiting value. Thus, most of the observations can be readily explained in terms of the dielectric properties of the minerals and saturating water, with the single exception being the large dielectric enhancement at low frequencies.

The Low-Frequency Dielectric Constant

The low-frequency dielectric behavior is illustrated by the data for a brine-saturated marble shown in figure IX.20a. Enhancement of κ' depends on the conductivity of the solution σ_w. As the conductivity

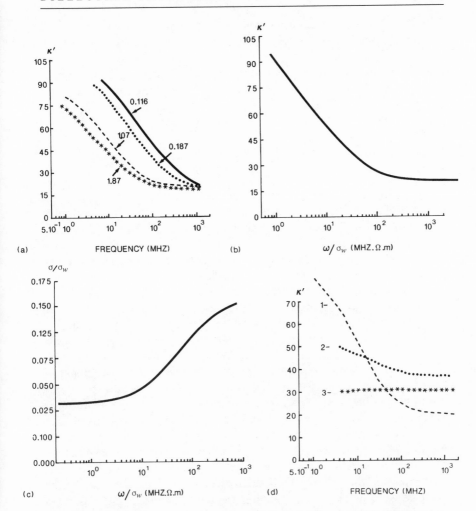

Fig. IX.20. (*a*) Dielectric constant κ' as a function of frequency for Whitestone marble saturated with water of four different resistivities: $\rho = 1.87\ \Omega\text{m}$; $\rho = 1.07$ Ωm; $\rho = 0.187\ \Omega\text{m}$; $\rho = 0.116\ \Omega\text{m}$. (*b*) Dielectric constant κ' as a function of ω/σ_w, where σ_w is the conductivity of the solution at 1 kHz. (*c*) Normalized effective conductivity σ/σ_w as a function of ω/σ_w. (*d*) Dielectric constant κ' as a function of frequency for 1—Whitestone marble; 2—powdered Whitestone marble; 3—powdered Vermont marble. Samples saturated with water of resistivity 1.07 Ωm. (After Kenyon, W. E., 1984, *J. Appl. Phys.* 55, 8:3153–59.)

increases the dielectric dispersion occurs at higher frequencies. All effects of salinity disappear at frequencies greater than 100 MHz. When this data is replotted as a function of ω/σ_w (instead of ω), all four curves collapse onto a single curve (fig. IX.20b). This scaling also applies for the effective conductivity (fig. IX.20c) and is consistent with our previous analysis. Earlier we have seen that the effective dielectric constant has the form

$$\kappa^* = \kappa_s F\{\kappa_w/\kappa_s\} = \kappa_s' F\{(\kappa_w'/\kappa_s')(1 - i\sigma_w/\omega\kappa_w'\varepsilon_o)\}$$

where F depends only on microstructure. For the relatively low salt concentrations used in this experiment, variations in κ_w' are negligible as compared to variations in σ_w. Therefore, the real part of F (effective dielectric constant) and imaginary part of F (effective conductivity) only depend on the single variable $(\sigma_w/\omega\kappa_w'\varepsilon_o)$. This is clearly supported by the results of the experiment (figs. IX.20b, c).

There still remains the important question "what aspect of the microstructure is responsible for the low-frequency enhancement of κ'?" A possible answer to this question is suggested by the results of several related experiments. Measurements were made on two additional samples of material similar to Whitestone marble, but with very different microstructures. The first was obtained by disaggregating (powdering) Whitestone marble, while the second was a powdered Vermont marble. Figure IX.20d allows a comparison of the Whitestone marble (at room temperature) with the two powdered samples. The powdered Whitestone marble shows less dispersion than the original sample, while the powdered Vermont marble shows no dispersion at all. Due to variations in porosity, the high frequency values are not equal. To determine the differences in microstructure, micrographs of the dry samples were made with a scanning electron microscope after impregnation with epoxy. The primary difference between samples was found to be the number of platy grains (grains with two long and one short dimensions); Whitestone marble having many, powdered Whitestone a few, and powdered Vermont marble almost none.

Platy grains can produce large polarizations. When a field perpendicular to a platy grain is applied, electrical charges of opposite sign can accumulate on opposite sides of a grain resulting in a large polarization. At the same time, the platy shape impedes conductive transport of the charge and lowers the conductivity. These observations are supported by measurements (table IX.3).

Archie's exponent m was obtained from the relation $\sigma = \sigma_w \phi^m$. The value $m = 1.5$ corresponds to spherical grains. Increasing values of m

Table IX.3. MICROSTRUCTURAL EFFECTS ON THE ELECTRICAL PROPERTIES OF ROCKS

Sample	Porosity ϕ	Rock Conductivity at 0.5 MHz $\sigma_{eff}(S / m)$	Archie Exponent (m)	Dielectric Dispersion
Whitestone marble (Intact)	0.30	0.055	2.3	High
Whitestone marble (powder)	0.55	0.30	1.9	Medium
Vermont marble (powder)	0.42	0.26	1.5	Small

Source: After Kenyon, W. E. 1984, *J. Appl. Phys.* 55, 8:3153–59.

correspond to progressively more oblate (disk-shaped) grains and with increasing dielectric dispersion.

Determination of Water Saturation

One very important application of the dielectric response of mixtures is in oil exploration. We have seen in chapter VIII that the effective conductivity can be used to estimate water saturation (and oil saturation if the porosity is known). This was due to the extremely large contrast in conductivities between water and mineral. Although the ratio of dielectric constant of water (80) to that of oil (2) is only 40, the presence of water should still measurably increase the dielectric constant of a rock.

The choice of measurement frequency is crucial. ω should be above the frequency domain of dielectric enhancement due to geometrical or charge layer effects. Thus ω should be > 20–200 MHz, with the exact lower limit depending on salinity (fig. IX.20). But ω must be lower than the dipolar relaxation frequency, $\omega < 10{,}000$ MHz, above which the dielectric constant of water decreases rapidly (fig. IX.6). Dielectric constant data in figure IX.21 was obtained at 500 MHz for a water-saturated sandstone and shows how κ' varies as a function of porosity ϕ. κ' decreases smoothly with decreasing ϕ and approaches a value of ~ 4.7 (value of amorphous silica) at $\phi = 0$. The experimental data can be represented by the empirical relation

$$\kappa^* = \left[x_1(\kappa_1)^s + \cdots + x_n(\kappa_n)^s \right]^{1/s} \text{ with } s = 1/2.$$

These results lead naturally to a second question: "Is there any information on the distribution or connectivity of porosity from these

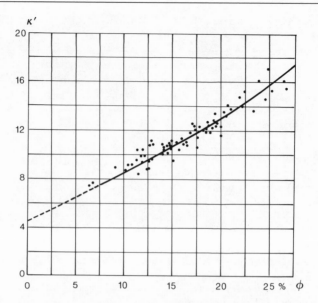

Fig. IX.21. Dielectric constant κ' of fresh water saturated sandstones as a function of porosity ϕ. Frequency = 500 MHz. (After Poley, J. P., J. J. Nooteboom, and P. J. deWaal, 1978, *The Log Analyst* [May–June], p. 8.)

measurements?" To find the answer Poley et al. (1978) measured κ' for these same samples after they had been partly saturated with oil. Saturation was accomplished in two ways, simulating both oil-wet and water-wet conductions. The two methods of saturation yield different spatial distributions of water within the pores for equal saturations. Figure IX.22 shows that there is no significant difference in κ' between the two groups of measurements. Thus, measurements of κ' in this frequency range indicate the amount of water saturation but not its distribution within the rock. Because κ' of oil (2.2) is comparable to that of quartz (4.9), replacing water by oil is almost equivalent to replacing water by quartz (decreasing porosity). Thus figure IX.22 will be almost identical to figure IX.21, if the axis of saturations S_w in (IX.22) is expressed as water fraction, that is, $0.22 \times S_w$ = (porosity) × (saturation).

When the preceding methods are applied to making measurements of water content in a well, several other considerations must be taken into account. There exist a number of electromagnetic tools operating in the frequency range of 20 MHz to 1.1 GHz. This choice of frequency range is based on competing considerations related to tool design, as well as

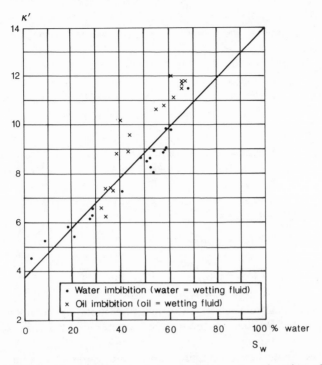

Fig. IX.22. Dielectric constant κ' of clean sandstones as a function of water saturation S_w. (Porosity = 22%; Frequency = 250 MHz). (After Poley, J. P., J. J. Nooteboom, and P. J. deWaal, 1978, *The Log Analyst* [May–June], p. 8.)

the electrical properties of the fluid-bearing rock. Three considerations are particularly important:

(a) *Attenuation:* Attenuation is determined by the loss tangent, tan $\delta = (\kappa'' + \sigma/\omega\varepsilon_o)/\kappa'$. At low frequencies attenuation is stronger due to conduction currents and varies as ω^{-1}. At higher frequencies, the dipolar term κ'' is important. Examining the curve IX.19b suggests that a good compromise for minimizing attenuation is a frequency around 1 GHz = 10^9 Hz.

(b) *Salinity:* The dielectric constant depends on salinity at lower frequencies (fig. IX.20a), but for $\omega \geq 1$ GHz it is independent of salinity. In the subsurface there will be no control over brine salinity. Thus to relate dielectric measurements to water content, without any independent salinity measurement, a choice of frequency $\omega \geq 1$ GHz is desirable.

(c) Volume of Investigation: The volume probed by a tool is determined by tool geometry, frequency range, and medium properties. At 20 MHz the wavelength is typically $\lambda \sim 10$ m (fig. IX.3) while at 1 GHz it is $\lambda \sim 0.1$ m. Depth of penetration $1/\beta$ increases as ω decreases: penetration is approximately 10–50 meters at 20 MHz and 1–10 meters at 1 GHz, depending on rock properties. Exploration at 20 MHz allows a much larger rock volume to be probed than at 1 GHz.

These considerations illustrate the compromises made in choosing a frequency for dielectric tool design. Tools operating at or close to 1 GHz operate in the low-loss region where the salinity effects are negligible. Dielectric logging tools that operate at lower frequencies (20–50 MHz) try to maximize the depth of penetration while still being in the propagation regime.

Effective Medium Approximation

Several analytical relations expressing κ^* as a function of the individual components of the medium have been presented in the preceding paragraphs. These relations were all empirical. It is also possible to obtain theoretical relationships through the effective medium approach.

Theories of disordered media approach the problem of calculating effective properties by considering an "average" homogeneous medium with dielectric constant κ_o^* and treating the grain-to-grain fluctuations in properties as perturbations on this homogeneous medium. Conceptually, the rock is divided into grains or cells, each with a homogeneous dielectric constant κ_i and then $(\kappa_o^* - \kappa_i)$ is considered a perturbation due to changing mineralogy.

For simplicity, let us assume that the grains are spherical. The first problem to consider is how an isolated dielectric sphere, of constant κ, when immersed in a homogeneous medium κ_o^* across which there is a constant field $\vec{E}_o$, will change that field. One can show that for this case the total electric field $\vec{E}$ inside the sphere becomes:

$$\vec{E} = \vec{E}_o + \vec{E}'$$

where $\vec{E}'$ is the field due to the sphere (the "perturbation"). The results of the calculation show (see problem X.d)

$$\vec{E} = \frac{3\kappa_o^*}{\kappa + 2\kappa_o^*}\vec{E}_o$$

$$\vec{P} = 3\varepsilon_o \frac{\kappa - \kappa_o^*}{\kappa + 2\kappa_o^*}\vec{E}_o.$$

These results for a single sphere can be extended to a distribution of spheres of volume fractions $x_1, x_2, \ldots x_i \ldots$ and constants $\kappa_1, \kappa_2 \ldots \kappa_i \ldots$, with the assumption that the interactions between spheres are negligible. One can also determine the quantities:

$$\langle \vec{E} \rangle = \langle \frac{3\kappa_o^*}{\kappa + 2\kappa_o^*} \rangle \vec{E}_o$$

$$\langle \vec{P} \rangle = 3\varepsilon_o \langle \frac{\kappa - \kappa_o^*}{\kappa + 2\kappa_o^*} \rangle \vec{E}_o$$

$$\langle \vec{D} \rangle = \varepsilon_o \langle \vec{E} \rangle + \langle \vec{P} \rangle$$

where the brackets are averages over the component volume fractions. Using the defining relation, IX.14 yields

$$\langle \vec{D} \rangle = \varepsilon_{\text{eff}} \langle \vec{E} \rangle = \kappa^* \varepsilon_o \langle \vec{E} \rangle.$$

It thus follows:

$$\kappa^* = \kappa_o^* \left[1 + 2 \sum_i x_i \frac{\kappa_i - \kappa_o^*}{\kappa_i + 2\kappa_o^*} \right] \left[1 - \sum_i x_i \frac{\kappa_i - \kappa_o^*}{\kappa_i + 2\kappa_o^*} \right]^{-1} \qquad \text{IX.15}$$

where x_i is the volume fraction of the i-th component. The theory is also applicable to alternating fields, if the wavelength is larger than the dimensions of the averaging volume. With this restriction the dielectric constants in equation IX.15 can all be frequency dependent.

As an illustration, consider a simple two-component system where a volume fraction x ($x \ll 1$) of material of constant κ_s is imbedded into a material with constant κ_w. The particles of material κ_s are assumed to be spheres. One can use equation IX.15 with the approximation that $\kappa_o^* = \kappa_w$. This is the *non-self-consistent* approximation.

$$\frac{\kappa^* - \kappa_w}{\kappa^* + 2\kappa_w} = x \frac{\kappa_s - \kappa_w}{\kappa_s + 2\kappa_w}.$$

This classical result is known as the generalized Clausius-Mossotti or Lorentz-Lorentz relation. Here κ_w is the continuous phase, and the phase κ_s only appears as dispersed grains. If κ_w represents water with a finite conductivity and κ_s a non-conducting mineral of volume fraction $1 - \phi$, the low-frequency limit will show a finite conductivity σ for the mixture:

$$\frac{\sigma}{\sigma_w} = \frac{2\phi}{3 - \phi}.$$

If the materials are interchanged, water being imbedded as isolated pores within a mineral matrix, the low-frequency conductivity would be zero. This latter case would be appropriate for modeling the dielectric response of a system where conductive minerals are dispersed within an insulating matrix.

The *self-consistent approximation*, first introduced in chapter III, consists of setting $\kappa^* = \kappa_o^*$. Equation IX.15 then simplifies to the result:

$$\sum_i x_i \left[\frac{\kappa_i - \kappa^*}{\kappa_i + 2\kappa^*} \right] = 0. \qquad\qquad \text{IX.16}$$

This approximation treats all components in an equal fashion (symmetrically). In practice it is usually in better agreement with experiment than the non-self-consistent approximation. For a mixture of conducting water and non-conducting rock, the result at low frequencies becomes:

$$\sigma = \left(\frac{3\phi - 1}{2} \right) \sigma_w \qquad \text{for } \phi > 1/3$$

$$\sigma = 0 \qquad \text{for } \phi \le 1/3.$$

The value of $\phi = 1/3$ corresponds to a percolation threshold.

Sedimentary rocks normally do not exhibit any percolation threshold, a result that is expected for rocks formed as granular aggregates. Thus the self-consistent approximation does not correctly model the low-frequency conductivity which is $\sigma/\sigma_w \sim \phi^m$. The non-self-consistent model has the correct topology (grains immersed in water), but is only accurate at large porosities (low rock-volume fractions $1 - \phi$).

A *self-similar* effective medium theory was developed by Sen et al. (1981) to overcome these difficulties. The basic idea of this approach is to consider a sequence of effective mediums, each succeeding medium being derived from the preceding one by a further addition of spherical grains. The beginning phase is pure water. At each stage j, a small volume fraction of spherical grains are immersed into the previous medium of constant $\kappa^{*(j)}$, and the new medium value $\kappa^{*(j+1)}$ is computed through the non-self-consistent model. This process is repeated until the desired volume fraction (porosity) is reached. For a rock composed of water and minerals, this model yields the result:

$$\phi = \left(\frac{\kappa_w}{\kappa^*} \right)^{1/3} \left(\frac{\kappa^* - \kappa_s}{\kappa_w - \kappa_s} \right). \qquad\qquad \text{IX.17}$$

A prime result of this model is that the pore space is always percolating! At low frequencies ($\sigma_w \gg \varepsilon_o \omega \kappa_w'$), conductivity has the Archie form with an exponent of $3/2$.

$$\frac{\sigma}{\sigma_w} = \phi^{3/2}.$$

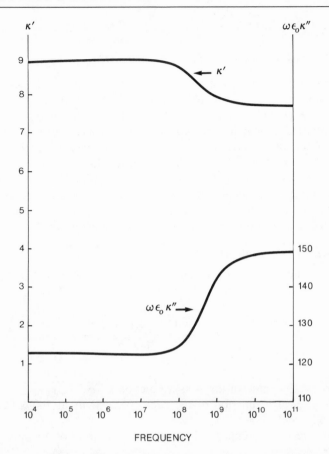

Fig. IX.23. Variation of κ' and $\omega\varepsilon_o\kappa''$ for an assembly of fused glass beads. $\phi = 0.009$, $\sigma_w = 4.55$ S/m. Self-similar model calculation. (After Sen, P. N., C. Scala, and M. H. Cohen, 1981, A Self-Similar Model for Sedimentary Rocks With Application to the Dielectric Constant of Fused Glass Beads, *Geophysics* 46:781–95.)

Figure IX.23 shows the predicted frequency behavior of κ' for a glass bead pack saturated with water. κ' has a typical Debye-type behavior. There is no low-frequency enhancement, which is consistent with the spherical-grain geometry.

The preceding theories are currently utilized for computing κ^* and for interpreting experimental measurements. We note in conclusion that a successful theory must be capable of predicting the experimentally determined variations of κ^* with both ϕ and ω. This can only be accomplished if a theory judiciously incorporates the important characteristics of rock microstructure.

PROBLEMS

IX.a. Cole-Cole Diagrams

(1) Assume that the response of a rock is analogous to the behavior of a circuit with a resistance R and capacitance C in parallel. Show that in the complex impedance plane (Z', Z'') the circuit is represented as a semicircle. Calculate the frequency value which corresponds to the top of the semicircle. Show that in the complex admittance plane (A', A''), one obtains a semi-infinite vertical line.

(2) In reality one observes, in the plane of complex admittances, a semi-infinite line inclined at an angle α. Derive how this result would appear in the complex impedance plane. To what aspect of the porous medium properties can this result be attributed?

IX.b. Depth of Penetration

Show that the depth of penetration of electromagnetic waves in a rock is given by

$$\frac{1}{\beta} = \frac{1}{\omega\sqrt{\mu_o \varepsilon'}} \frac{[1 - \tan^2(\delta/2)]^{1/2}}{\tan(\delta/2)}.$$

(1) Express the value of $\tan \delta$ as a function of $\varepsilon', \varepsilon'', \sigma, \omega$.
(2) Plot $1/\beta$ as a function of the rock conductivity at 100 MHz. Take $\varepsilon' = 5\, \varepsilon_o$ and $\varepsilon'' = 0.1\, \varepsilon_o$.
(3) What can one deduce about the possibilities of prospecting by radar?

IX.c. Interface Polarization

Consider an assemblage of two layers of thickness, d_1 and d_2. The first layer is a dielectric of constant κ_1 and zero conductivity. The second layer is a conductor of conductivity σ_2 and zero dielectric constant. Show that the composite system has the equivalent dielectric constant

$$\kappa^* = \frac{\kappa_1\left(1 + \dfrac{d_1}{d_2}\right)}{1 - i\kappa_1\left(\dfrac{\omega\varepsilon_o}{\sigma}\right)\left(\dfrac{d_2}{d_1}\right)}.$$

Show that elevated values of κ^* can be obtained (Maxwell-Wagner mechanism).

REFERENCES

Cole, K. S., and R. H. Cole. 1941. Dispersion and Absorption in Dielectrics I. Alternating Current Characteristics. *Journal of Chemical Physics*, 9:341–51.

Debye, P. 1929. *Polar Molecules*. Chap. 5, pp. 77–108. New York: Dover.

Keller, G. 1987. Rock and Mineral Properties, in *Electromagnetic Methods in Applied Geophysics-Theory*, vol. 1, edited by M. N. Nabighian, Society of Exploration Geophysicists, Tulsa, Okla.

Olhoeft, G. R. 1979. Tables of Room Temperature Electrical Properties for Selected Rocks and Minerals with Dielectric Permittivity Statistics. *U.S. Geological Survey Open File Report 79–993*.

———. 1981. Electrical Properties of Rocks, in vol. II-2 *Physical Properties of Rocks and Minerals*, McGraw Hill/Cindas Data Series on Material Properties, edited by Y. S. Touloukian and C. Y. Ho.

Parkhomenko, E. I. 1967. *Electrical Properties of Rocks*. New York: Plenum Press, pp. 24–29.

Von Hippel, A. R. 1954. *Dielectrics and Waves*. New York: John Wiley and Sons, pp. 21–25.

X. Thermal Conductivity

THERMAL PROPERTIES fall into two general classes: those related to equilibrium properties and those associated with non-equilibrium transport of energy. In the first group one finds the specific heat and the coefficient of thermal expansion. These parameters allow us to relate internal energy and volume changes to changes in temperature. In the second group one finds the coefficient of thermal conductivity. This coefficient is a measure of how rapidly heat is transported from one point in a medium to another.

In this chapter we will primarily focus on the thermal conductivity and specific heat, which are the two important parameters appearing in equations describing heat transport. Knowledge of these parameters is important in a number of applications such as the extraction of geothermal energy, the storage of nuclear wastes, and for predicting the thermal evolution of igneous intrusions. Our predictions can only be as accurate as our knowledge of the thermal properties of the rocks that we are modeling.

LAW OF HEAT CONDUCTION

Conduction and Convection: Fourier's Law

Convection is one of the principal mechanisms of energy transport. The transport of energy by convection is proportional to the energy content of the fluid cT (c is the specific heat of the fluid in cal cm^{-3} K^{-1} and T is the temperature in degrees Kelvin) and the volume flux of fluid $\vec{q}$ (cm sec^{-1}).

$$\text{Convective Flux} = cT\vec{q} \quad (\text{cal cm}^{-2} \text{ sec}^{-1}).$$

The velocity $\vec{q}$ is Darcy's velocity which is proportional to the rock permeability and pressure gradient (equ. V.1: Darcy's law). Because there appears to be a finite permeability in most rocks to a depth of 10 kilometers or so, convection is a major mechanism for energy transport in the crust.

Transport of energy can also take place when there is no net transport of mass. In these cases the energy is transported by lattice vibrations or through the emission and absorption of radiation. These processes are referred to as thermal conduction and the heat flux is described by Fourier's law:

$$\text{Conductive Flux} = J_i = -\lambda_{ij}\frac{\partial T}{\partial x_j}. \qquad \text{X.1}$$

Here λ_{ij} is the thermal conductivity tensor. Most rock-forming minerals such as quartz and calcite are anisotropic and thus λ_{ij} is in general anisotropic. In the simpler case of an isotropic rock, $\lambda_{ij} = \lambda\delta_{ij}$ (where $\delta_{ij} = 0$ if $i \neq j$ and $\delta_{ij} = 1$ if $i = j$). In this case Fourier's law becomes, for a temperature gradient in the x-direction,

$$J_x = -\lambda\frac{dT}{dx}.$$

Fourier's law is analogous to the laws of Ohm and Darcy. It states that the heat flux is proportional to the temperature gradient (the "force" in the sense of irreversible thermodynamics).

Thermal Conductivity and Diffusivity

The thermal conductivity of an isotropic mineral is most easily determined by measuring the steady-state heat flux J that traverses a sample of thickness Δx, across which there is a temperature difference ΔT. If the sample is small and homogeneous, the temperature gradient within the sample will be constant and have the value $\Delta T/\Delta x$.

$$\lambda = -\frac{J}{\Delta T/\Delta x}.$$

J is expressed in units of (Watts m^{-2}) or (cal cm^{-2} sec^{-1}). The conductivity λ is expressed in units of (Watts m^{-1}°C^{-1}) or (cal cm^{-1} sec^{-1}°C^{-1}). Conversion factors between these units are listed below.

$$1 \text{ Watt m}^{-2} = 2.388 \times 10^{-5}\text{cal cm}^{-2}\text{ sec}^{-1}$$
$$1 \text{ Watt m}^{-1}°C^{-1} = 2.388 \times 10^{-3}\text{cal cm}^{-1}\text{ sec}^{-1}°C^{-1}.$$

The thermal evolution of a rock is described through the conservation of energy equation. Consider the energy balance in a small element of thickness δx (fig. X.1), for a time interval δt:

$$C^*\delta T\,\delta x = [J(x) - J(x + \delta x)]\delta t = -\delta J(x)\delta t.$$

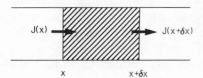

J(x)   J(x+δx)

x   x+δx

Fig. X.1. Energy conservation, over a time interval δt, for a 1-dimensional model. $C^*\delta T\,\delta x = (J(x) - J(x + \delta x))\delta t$.

Because the inflow of heat $J(x)$ at x is not equal to the outflow at $x + \delta x$, there will be an increase in temperature δT during the time δt. Thus in one dimension, in the absence of heat sources:

$$C^* \frac{\partial T}{\partial t} = -\frac{\partial J}{\partial x}.$$

Generalizing to three dimensions, with the presence of heat sources H and convection, this equation becomes:

$$C^* \frac{\partial T}{\partial t} + \vec{\nabla} \cdot \vec{J} = H. \qquad \text{X.2}$$

Here C^* is the average specific heat of the porous rock (possibly saturated with fluid) and J is the total heat flux (conductive and convective). The term $\left(C^* \dfrac{\partial T}{\partial t} \right)$ represents the change in the thermal energy per unit time, and the term $(\vec{\nabla} \cdot \vec{J})$ represents the inflow of heat per unit time (fig. X.1). H is a source term which describes the internal energy generated interior to the volume segment δx per unit volume and time (for example, the heat produced by radioactive elements contained in a granitic rock). For the case of an isotropic rock with no heat sources, sinks ($H = 0$), or convection ($\vec{q} = 0$), equations X.2 and X.1 yield:

$$\frac{\partial T}{\partial t} = \alpha \nabla^2 T. \qquad \text{X.3}$$

The coefficient $\alpha = \lambda / C^*$ is the thermal diffusivity, expressed in m^2 sec^{-1} or $\text{cm}^2 \text{ sec}^{-1}$. Typical values for rocks are: $\lambda = 3$ Watts $\text{m}^{-1}{}^\circ\text{C}^{-1}$ and $C^* = 2 \cdot 10^6$ J m^{-3}, thus $\alpha = 10^{-6}$ m^2 sec^{-1}. If transport of heat takes place over a distance h, then the ratio $\tau \equiv h^2/\alpha$ is the characteristic time for this transport to take place. For times $t \gg \tau = h^2/\alpha$, the system will be very close to *steady state* $\left(\dfrac{\partial T}{\partial t} \approx 0 \right)$. The exact solutions of X.3 for different conditions and limits can be expressed in terms of this time constant τ.

THERMAL CONDUCTIVITY OF MINERALS

The thermal conductivity of rocks is principally determined by the thermal conductivities of the minerals and saturating fluids that constitute the rock. Microstructure plays a minor role and the same is true for percolation affects. Percolation is not important in conduction, as it is in the convective heat transport (hydrothermal for example) where the permeability is an essential parameter. Before examining the thermal conductivity of rocks, we first consider the termal conductivity of minerals.

Conductivities and Specific Heats

Mineral thermal conductivities are often strongly anisotropic. For the case of quartz (at $T = 0°C$ and $P = 1$ atm) the conductivity parallel to the optic axis is 11.39 W m^{-1}°C^{-1}, but only 6.46 W m^{-1}°C^{-1} perpendicular to this axis. Because rocks are aggregates of mineral grains, anisotropy often disappears due to the random orientation of the grains. But, if the rock sample is very small and approaches the size of individual grains, anisotropy can be very strong. At the opposite extreme, very large rock masses can also be anisotropic due to bedding and schistosity.

In the temperature range found in the earth's upper crust (300°K to 600°K), the thermal conductivities of most crustal rocks and minerals range between (1 and 10 W m^{-1}°C^{-1}). Figure X.2 shows that quartz and calcite exhibit a strong and complicated variation at lower temperatures (T < 300°K), but at high temperatures (T > 300°K) the variation is much smaller. Conductivities of other rocks and minerals generally fall below that of calcite and quartz over the whole temperature range.

Most laboratory measurements are made under steady-state conditions. One compares the thermal conductivity of the sample to be measured (λ) with a standard reference of conductivity (λ_1). The sample and reference are disk shaped, held together, and the heat flux is parallel to the axes:

$$J = \lambda_1 \frac{\Delta T_1}{\Delta x_1} = \lambda \frac{\Delta T}{\Delta x}.$$

ΔT_1 and ΔT are the temperature differences across the reference and the material of unknown conductivity, and Δx_1 and Δx are their respective thicknesses.

In addition to this steady-state method, transient methods have also been utilized and offer several practical advantages. To measure the thermal conductivity of a polycrystalline aggregate, a specimen is pulverized into a fine powder and mixed with distilled water of known conductivity λ_w. Small-diameter grain sizes are preferred so as to lower the permeability of the powder and prevent any convective heat transfer. Another reason for the small grain size is that the mixture must be macroscopically homogeneous and isotropic. A needle point probe (heater) is inserted into the mixture and a constant source of heat H is supplied. Because the mixture is analyzed by the theory of thermal

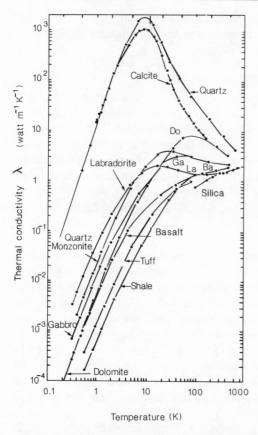

Fig. X.2. Thermal conductivities of different crustal rocks, compared to minerals α-quartz, calcite, and various silicates: Do = Dolomite; La = Labradorite (An 65.5); Ga = Gabbro, Ba = Basalt. (After Pohl, R. O., 1985, Heat Transport in Rocks, Glasses, and Disordered Crystals, in *Transport and Relaxation in Random Materials*, ed. J. Klafter, R. J. Rubin, M. F. Shlesinger, World Scientific Pub. Co. Pte. Ltd., Philadelphia, Pa.)

conduction in an isotropic and homogeneous media, the size of the mineral grains must be much smaller than the sample size (typically ~ 100 cm^3) and the needle probe. The temperature rise ΔT is then measured as a function of time t at a fixed distance R away from the probe (Von Herzen and Maxwell, 1959). If the needle is "sufficiently" long, the problem has cylindrical symmetry and can be analyzed by the well-known theoretical solution to equation X.2. The effective thermal conductivity λ_{eff} is found from:

$$\lambda_{\text{eff}} = \frac{H}{4\pi\,\Delta T} \ln\left(\frac{t}{t_o}\right)$$

where $\Delta T(t)$ is the temperature rise at R at time t, and t_o is a constant. The conductivity λ_s of the solid (powder) is obtained from the value λ_{eff} (measured) and λ_w (known), utilizing an effective medium model. One example is:

$$\lambda_{\text{eff}} = \frac{1}{2}(\lambda' + \lambda'').$$

Here λ' and λ'' are the upper and lower Hashin-Shtrikman bounds (chap. III) for a two-component mixture. The first constituent is water (conductivity λ_w, volume fraction ϕ) and the second a polycrystalline solid (conductivity λ_s, volume fraction $1 - \phi$). Letting $\delta\lambda = \lambda_W - \lambda_S$

$$\lambda' = \lambda_s\left[1 + \frac{3\phi\,\delta\lambda}{3\lambda_s + (1 - \phi)\delta\lambda}\right]$$

$$\lambda'' = \lambda_w\left[1 - \frac{3(1 - \phi)\delta\lambda}{3\lambda_w - \phi\delta\lambda}\right].$$

Other effective medium theories can also be utilized for computing λ_s from λ_{eff}. The value of λ_s obtained in this manner will depend on the effective medium model chosen. Note that the above equations also appear following VIII.11. The results derived for the electrical conductivity are directly applicable here, due to the formal analogy between heat and electrical current flow as discussed in chapter III.

Table X.1 contains experimental values of specific heats and conductivities of mono-mineralic aggregates. The values presented for anisotropic minerals is an average over all crystal orientations. More extensive compilations of rock and mineral thermal properties can be found in the reviews by Roy et al. (1981), Cermak and Ryback (1982), and Robertson (1988).

It is important to emphasize that even when the measurement precision is very high (less than 1%), there can be variations in conductivities of adjacent samples of the same rock as large as 30%. These variations result from rock heterogeneity due to varying composition and porosity. For this reason the general philosophy is to design the measuring apparatus to give rapid results to within 5% accuracy and to average the results to arrive at a representative value for the rock type.

Thermal Conductivity Models

It is useful to have a theoretical understanding of how λ varies and how it can be correlated to other properties. Thermal conductivity depends on many factors. In general the transport of energy takes place through lattice vibrations at low and average temperatures, and through radiation at very high temperatures. We will examine the second process later. For the conditions of the upper crust (0°C to 300°C and 1 bar to 2 kbars) the dominant transport process is through lattice vibrations. Because rock-forming minerals are dielectrics, electronic contributions are negligible.

Debye's model provides the principal results for understanding λ, in particular its inverse dependence on temperature $\lambda \sim \dfrac{1}{T(K°)}$ that is

Table X.1. THERMAL CONDUCTIVITIES AND SPECIFIC HEATS C_p, OF
MINERAL AGGREGATES

Mineral	Formula	λ (Watt / m-°C) at 25°C	c_p (kJ / kg-°C) at 0°C
Carbonates			
Calcite	$CaCO_3$	3.57	0.793
Aragonite	$CaCO_3$	2.23	0.78
Magnesite	$MgCO_3$	5.83	0.864
Dolomite	$CaMg(CO_3)_2$	5.50	0.93 (at 60°C)
Silicates			
α-Quartz	α-SiO_2	7.69	0.698
Amorphous Silica	SiO_2	1.36	0.70
Olivines			
Forsterite (Fo)	$Fo_{98}Fa_2$	5.06	
Fayalite (Fa)	Fo_4Fa_{96}	3.16	0.55
Pyroxenes			
Enstatite $En_{98}Fs_2$		4.34	0.80 (at 60°C)
Bronzite $En_{78}Fs_{22}$		4.16	0.752
Augite	$CaMgFeSi_3O_9$	3.82	
Diopside	$CaMgSi_2O_6$	5.02	0.69
Wollastonite	$CaSiO_3$	4.03	0.67
Amphiboles			
Antophyllite	$Mg_7Si_8O_{22}(OH)_2$	3.96	0.740
Tremolite	$Ca_2Mg_5Si_8O_{22}(OH)_2$	4.08	
Hornblende	$Ca_2Mg_3FeAlSi_7O_{22}(OH)_2$	2.88	
Glaucophane	$Na_2Mg_3Al_2Si_8O_{22}(OH)_2$	2.17	
Alkali Feldspar			
Microlcline		2.49	0.680
Orthoclase	$KAISi_3O_8$	2.31	0.61
Sanidine		1.65	
Plagioclase			
$Ab_{99}An_1$	Ab = $NaAlSi_3O_8$(Albite)	2.31	0.709
$Ab_{46}An_{54}$		1.53	0.70
Ab_4An_{96}	An = $CaAl_2Si_3O_8$(Anorthite)	1.68	0.70
Evaporites			
Halite	$NaCl$	6.10	
Anhydrite	$CaSO_4$	4.76	0.52

Source: After Horai, K., 1971. Thermal Conductivity of rock-forming minerals, *J. Geophysical Research* 76:1278–1308; and Goranson, J., 1942, in Birch, F. *Handbook of Physical Constants*, p. 228, Geol. Soc. America Spec. Paper no. 36.

observed above 25°C. When a crystal lattice is at a temperature T, the lattice vibrations can be separated into individual traveling waves (phonons in the nomenclature of quantum mechanics). These traveling waves can be viewed as quasi-particles of specific energy in the kinetic theory of gases. By analogy to the kinetic theory of gases, the thermal conductivity of a phonon "gas" is

$$\lambda = \left(\frac{1}{3}\right) CVL \qquad \text{X.4}$$

where C is the specific heat per unit volume of "gas," V is the average particle velocity, and L is their mean free path. L is the average distance a particle will travel between two successive collisions. Although equation X.4 is exceedingly simple and a gross approximation of reality, it describes the fundamental physical processes surprisingly well. X.4 has the form of a transport equation: CV represents the "convective" energy flux of a gas, and L is the distance of transport. If the sample length is $\Delta x = L$, then the transported energy flux is $J = \lambda \dfrac{\Delta T}{\Delta x} = \dfrac{1}{3}(CV)\Delta T$. But in general $\Delta x \gg L$ and thus $J \ll (CV)\Delta T$: the successive phonon collisions significantly limit the efficacy of energy transport.

The variation of λ results from variations of C, V, or L. At high temperatures, $C = 3Nk_B$ for all solids, where $k_B = 1.38 \times 10^{-16}$ ergs/deg is Boltzmann's constant and N is the number of atoms considered. For one mole of solid, $N = 6.03 \times 10^{23}$ atoms and $C = 25$ joules/mole-deg (6 cal/mole-deg). At low temperatures, C decreases rapidly and approaches zero (fig. X.3). The high- and low-temperature regimes are defined in relation to the Debye temperature T_D of a mineral, the magnitude of which only depends on the properties of the crystal. At high temperatures, $T > T_D$, there is sufficient thermal energy to excite all possible lattice waves (all degrees of freedom, $6/\text{atom} \times N$ atoms). Because each degree of freedom has thermal energy $\frac{1}{2}k_B T$, thus $C = 3Nk_B$, the variation of λ due to C is usually negligible for most common rock-forming minerals in the temperature range 0°–300°C.

The second term in equation X.4 is the average velocity V. This velocity is calculated in Debye's theory from the longitudinal (V_p) and shear (V_s) velocities of a crystal:

$$\frac{1}{V^3} = \frac{1}{3}\left[\frac{1}{V_p^3} + \frac{2}{V_s^3}\right].$$

This relation for the average velocity takes into account that there are twice as many transverse vibrational degrees of freedom as there are

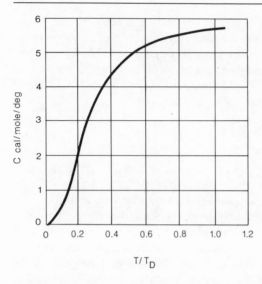

Fig. X.3. Specific heat of solids in the Debye approximation.

longitudinal. The velocities are not very sensitive to changes in temperature.

Variations in L arise from two types of effects: collisions between phonons and collisions with crystal lattice imperfections. For $T > T_D$, collisions betwen phonons result in the dependence $L \sim 1/T$. Thus L becomes quite large at low temperatures. At ambient temperatures for example, $L = 40$ Å for α-quartz. Thus L is approximately 10 times longer than the bond length, but it is still much shorter than the grain dimensions. Lattice imperfections of diverse nature can also be present: point defects, dislocations, and grain boundaries can limit the size of L if their density is sufficiently great. A simple approximation for calculating the effective mean free path when several mechanisms are operating is:

$$\frac{1}{L} = \frac{1}{L_1} + \frac{1}{L_2} + \frac{1}{L_3} + \cdots$$

where L_n is the mean free path for process n. Thus the process with the shortest mean free path is the one that limits the conductivity. A lower bound for L is attained in amorphous substances which can be considered to be completely disordered. In this case L is of the order of the dimension of the unit cell (5 to 10 Å). This explains why the thermal conductivity of amorphous SiO_2 is approximately 6 times smaller than that of α-quartz (table X.1).

Conductivity also varies with pressure P: the increase in λ with P essentially reflecting the increase of the average velocity V. This varia-

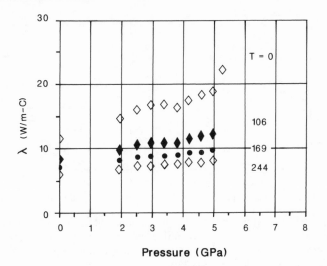

Fig. X.4. Thermal conductivity of quartz (parallel to the optic axis) as a function of pressure P and temperature T. (After Beck, A. E., D. M. Darbha, and H. H. Schloessin, 1978, Lattice Conductivities of Single-Crystal and Polycrystalline Materials at Mantle Pressures and Temperatures, *Physics of the Earth and Planetary Interiors* 17:35–53.)

tion is small and a linear approximation is sufficient:

$$\lambda \cong \lambda_o \left[1 + g\left(\frac{P}{K_s}\right) \right].$$

Here K_s is the mineral bulk modulus and g a dimensionless constant. Figure X.4 shows that, for the conditions of the Earth's crust, the effects of temperature are much more important than those of pressure. A pressure of 1 GPa = 10 kb corresponds to the average depth of the crust-mantle boundary (35 km).

Correlations with Other Physical Properties

The search for correlations between thermal conductivity and other physical properties is motivated by the fact that it is difficult to directly measure λ at high temperatures and pressures. The preceding kinetic model suggests that λ is proportional to the average acoustic velocity, specific heat, and mean free path of the phonons. Unfortunately, the mean free path is difficult to predict and is usually the dominant factor.

A good correlation between λ and A, the atomic mass, has been observed for simple solids (semiconductors). Similar correlations have also been observed for carbonates (fig. X.5). Although there is a

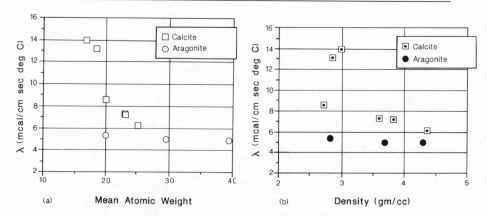

Fig. X.5. (*a*) Correlation between thermal conductivity and atomic mass (carbonates). (*b*) Correlation between thermal conductivity and density (carbonates). (After Horai, Ki-iti, 1971, Thermal Conductivity of Rock-Forming Minerals, *J. Geophysical Research* 76:1278–1308).

correlation, there is a clear difference between these two groups as seen by the markedly different slopes: -2.1 for the calcite group and -0.12 for the aragonite group. If the conductivity is plotted versus the mineral density, the correlations remain linear except for the two minerals containing magnesium (magnesite and dolomite).

A very good correlation is observed between λ and the compressional (V_p) and shear (V_s) velocities: $V = a + b\lambda$ (fig. X.6). The relative dispersion in data is attributable to the effects of lattice imperfections on L. This correlation can be understood through Debye's model.

Finally, a structural effect can be observed for silicates (fig. X.7). Due to the tetrahedral coordination of the covalent Si-O bond, the silicon and oxygen form a "rigid" tetrahedra $(SiO_4)^{-4}$, the arrangements of which determine the fundamental framework of the silicates. Due to the rigidity of the bond, we expect phonons to be transmitted more effectively through this framework, and thus the conductivity to increase as the coordination of tetrahedra between structures increases. This is what is observed. There is general increase in the average conductivity as the coordination of the silicon tetrahedra increases.

Radiative Conductivity

It has been observed for a long time that at very high temperatures the thermal conductivity increases with temperature. This effect is attributed to radiative transfer. Conceptually, radiative transfer is analogous to conduction but the energy is transported by electromagnetic

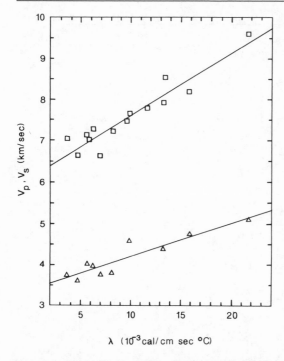

Fig. X.6. Correlation between thermal conductivity and velocities V_p and V_s (After Horai, Ki-iti, 1971, Thermal Conductivity of Rock-Forming Minerals, *J. Geophysical Research* 76:1278–1308).

waves (photons) rather than lattice vibrations (phonons). At a given temperature T there exists an equilibrium distribution of photons. If there is a temperature difference between two points in the solid, there will be a transfer of energy by photons, in addition to the energy transfer by phonons. Radiative transfer is proportional to the energy content of the electromagnetic radiation (which varies as T^3), the velocity of the waves in the solid, and the average mean free path. This results in the radiative component of the conductivity being proportional to T^3.

Lattice and radiative thermal transfer occur simultaneously as is shown in figure X.8. The apparent conductivity of $CaCO_3$ (measured at $P = 0.7$ GPa) follows a $1/T$ law, as is expected for conduction by phonons. Around 500°K, there occurs a deviation from linearity which is interpreted as the onset of radiative transfer. The radiative component λ_r can be isolated by subtracting the low-temperature $1/T$ behavior from the observed conductivity (fig. X.8b). The observed linearity of λ_r and T^3 supports the interpretation that this additional heat transfer is due to radiative transfer. Because the lattice contribution decreases with increasing temperature as $1/T$, while the radiative contribution increases as T^3, there is a minimum in the conductivity at around 800°K.

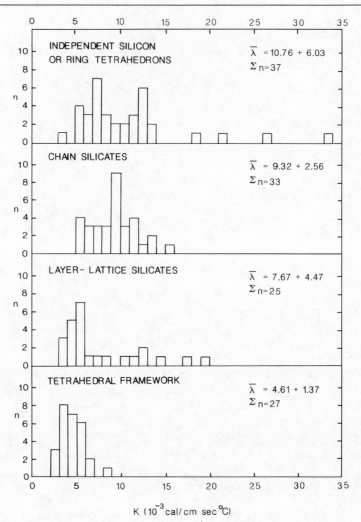

Fig. X.7. The effect of silicate structure on thermal conductivity. (After Horai, Ki-iti, 1971, Thermal Conductivity of Rock-Forming Minerals, *J. Geophysical Research* 76:1278–1308).

Birch and Clark (1940) were the first to suggest that radiative heat transfer might be an important mechanism in the Earth's mantle. Their extrapolations based on the T^3 law suggested very large values for λ_r at 1000°K and higher. Subsequent measurements have shown that increasing pressure and temperature also decrease the mean free path of radiation, thus limiting λ_r. It is now believed that the radiative compo-

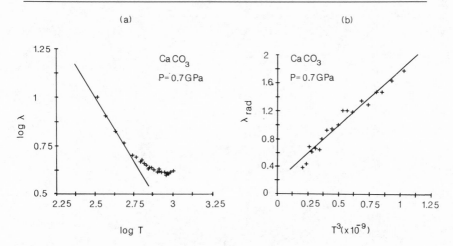

Fig. X.8. (*a*) Total thermal conductivity λ of $CaCO_3$ at $P = 0.7$ GPa. Notice the deviation from the $1/T$ law at 500°C due to radiative transfer. (*b*) Radiative conductivity λ_r of $CaCO_3$ at $P = 0.7$ GPa. (After MacPherson, W. R., and H. H. Schloessin, 1982, Lattice and Radiative Thermal Conductivity Variations Through High P, T Polymorphic Structure Transitions and Melting Points, *Physics of the Earth and Planetary Interiors* 29:58–68).

nent λ_r is small and does not exceed the lattice contribution under the conditions of the Earth's crust and mantle.

THERMAL CONDUCTIVITY OF ROCKS

One of the earliest syntheses of the thermal conductivity of rocks and minerals of the upper crust is that by Birch and Clark (1940). Their early measurements and discussion provided a broad overview of rock and mineral thermal properties in the temperature range 0 to 400°C, a part of which is reproduced in figure X.9. This data shows the observed variations for single crystals and polycrystalline sedimentary (9*a*) and igneous (9*b*) rocks. Conductivities decrease as T increases from 0 to 400°C, and although the range of variation is smaller at 400°C than at 0°C, it is still not negligible. Birch and Clark noted that feldspars, due to their lower conductivity, play an important role in determining the conductivity of igneous rocks: lower feldspar content (i.e., dunite, bronzitite, hypersthenite) implies higher thermal conductivity. Plagioclase feldspar anorthosite has one of the lowest conductivities and contrary to crystalline minerals, its conductivity increases slowly with temperature. This particular temperature dependence is more characteristic of solids with an amorphous structure and indicates a high degree of disorder.

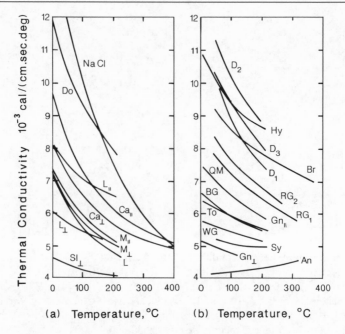

Fig. X.9. Thermal conductivity of rocks and minerals of the crust. (After Birch and Clark, 1940.) (*a*) Sedimentary rocks; (*b*) Igneous rocks.

Quartz on the contrary has one of the highest conductivities, and thus rock conductivity generally increases as the quartz content increases. In this section we will investigate quantitatively how the effective thermal conductivity of a rock is influenced by its mineralogy, amount and type of saturating fluids, and microstructure.

Conductivity of Aggregates

As discussed in chapter III, the close analogy between thermal and electrical conductivity (in the absence of surface conduction) allows us to directly apply the results of chapter VIII. The effective thermal conductivity of an aggregate material is determined by the relation between the average flux $\overline{J(r)}$ and the temperature gradient $\Delta T/\Delta x$

$$\overline{J(r)} = -\lambda_{\text{eff}} \frac{\Delta T}{\Delta x}.$$

The flux is averaged over the sample volume. The preceding definition is analogous to that utilized for determining λ for minerals (powdered). Note that minerals are 3 to 10 times more conductive than water. Thus thermal energy is conducted more efficiently through solids than water, contrary to what is observed for the electrical conductivity.

The effective thermal conductivity λ_{eff} depends on the conductivities of the mineral constituents λ_i, their volume fractions x_i, and microstructure. In the general case $\lambda_{eff} = F(x_i, \lambda_i)$ where F depends on the microscopic distribution. As for the dielectric constant, an example of a function F that is often utilized is the geometric average. For a system of n constituents with conductivities λ_i and volume fractions x_i:

$$\lambda_{eff} = \lambda_1^{x_1} \lambda_2^{x_2} \ldots \lambda_i^{x_i} \ldots \lambda_n^{x_n}.$$

Effective medium computations are also possible. Just as for the dielectric sphere case in chapter IX, it is illustrative to consider the problem of a homogeneous sphere of conductivity λ, placed in a homogeneous medium of conductivity λ_o^* (see prob. X.d). The calculation and results are analogous to those presented in chapter IX. Temperature plays the role of the electrical potential and the temperature gradient the role of the electric field. For the case of a distribution of solid ($\lambda = \lambda_s$) and liquid ($\lambda = \lambda_w$) spheres, one obtains for the effective conductivity:

$$\lambda_{eff} = -2\lambda_o^* + \cfrac{1}{\cfrac{\phi}{2\lambda_o^* + \lambda_f} + \cfrac{1-\phi}{2\lambda_o^* + \lambda_s}}.$$

ϕ is the porosity (assumed saturated with water). This result is exactly analogous to IX.15.

The Hashin-Shtrikman bounds for the case of a two-component aggregate have been previously presented. For the more complex case of m components, it is possible to define the bounds for an isotropic mixture by arranging the components in increasing magnitude: that is, $1 \leq i \leq m$, $\lambda_1 \leq \lambda_2 \cdots \leq \lambda_m$. Then the lower, λ_{min}, and upper, λ_{max}, are

$$\lambda_{min} = \lambda_1 + \frac{3\lambda_1 A_1}{3\lambda_1 - A_1}$$

$$\lambda_{max} = \lambda_m + \frac{3\lambda_m A_m}{3\lambda_m - A_m}$$

X.5

where $A_1 = \displaystyle\sum_{i=2}^{m} \frac{x_i}{(\lambda_i - \lambda_1)^{-1} + (3\lambda_1)^{-1}}$ and

$$A_m = \sum_{i=1}^{m-1} \frac{x_i}{(\lambda_i - \lambda_m)^{-1} + (3\lambda_m)^{-1}}.$$

Because the rock-forming minerals generally have conductivities within an order of magnitude of each other, the Hashin-Shtrikman bounds are quite narrow.

Table X.2. Composition and Conductivities of Minerals Comprising an Average Alkali Granite ($T = 0°C$)

Mineral	Volume Fraction	$\lambda (W\,m^{-1\circ}C^{-1})$
Amphibole	.10	2.52
Quartz	.26	7.69
Alkali feldspar	.64	2.31

Granites

The previous formulas allow us to determine the range of possible values for the thermal conductivity of a rock with the composition of an average alkali granite. Approximate compositions (in volume fractions) and conductivities are shown in table X.2.

The H.S. bounds are:

$$(\lambda_{min} = 3.20) \le \lambda_{eff} \le (3.45 = \lambda_{max})\,(W\,m^{-1\circ}C^{-1}).$$

A practical estimate for λ_{eff} is the average of the two bounds, $\lambda_{eff} = 3.32 \pm 0.125$. The uncertainty is very small (4%) considering that no information on microstructure, except for isotropy, was utilized. In "homogeneous" rock units, fluctuations in conductivity between closely spaced samples are often larger than 4%. From this viewpoint the H.S. bounds are quite narrow and provide a "good" estimate for the conductivity.

Because granites usually contain a small fracture porosity, we consider now what effect this porosity has on the previous bounds. Suppose that the fracture porosity is .02 and isotropically distributed. We can calculate the bounds for a four-component aggregate with the fourth component being the fluid in the fractures. If the saturating fluid is water (.565 W $m^{-1\circ}C^{-1}$), the H.S. bounds are:

$$2.87 \le \lambda_{eff} \le 3.39.$$

The average value is now lower and the bounds wider: $\lambda_{eff} = 3.13 \pm 0.26$ W $m^{-1\circ}C^{-1}$. By incorporating additional information, these bounds can be improved. For example, if the average fracture length is much larger than the average grain (but still much smaller than the sample size), we can replace the amphibole-quartz-feldspar aggregate by a homogeneous equivalent medium, a non-porous "effective granite" with a conductivity 3.20 (lower bound) or 3.45 (upper bound). We can repeat the process and introduce a fracture porosity of .02 into both end members and compute the bounds for both. If the granite has the upper value of 3.45, then the fractured system has bounds $3.3 \le \lambda_{eff} \le 3.37$. If the granite has the lower value 3.20, then the bounds are $3.07 \leq \lambda_{\text{eff}} \leq 3.13$. This implies that the fractured system would be bounded by the upper bound of the higher value and the lower bound of the lower value.

$$3.07 \leq \lambda_{\text{eff}} \leq 3.37.$$

As before, the average of the bounds yields $\lambda_{\text{eff}} = 3.22 \pm 0.15$. Note that the uncertainty has been reduced by incorporating this additional information about the microstructure.

Thermal conductivity varies with depth as the pressure and temperature increase. Because the porosity of granites is very low, the effect of pressure is negligible as compared to that of temperature. Figure X.10

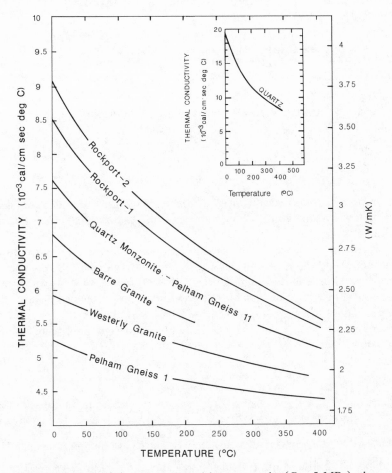

Fig. X.10. Variation $\lambda(T)$ for a series of igneous rocks ($P = 5$ MPa): the temperature variation is principally determined by quartz, with the differences between rocks reflecting their quartz content.

shows the temperature dependence of the conductivity for various igneous rocks. Conductivity differences between rocks is essentially controlled by their quartz content.

Role of Porosity, Fluids, and Fractures

Before examining the case of sedimentary rocks, it is useful to consider the effects of porosity and saturating fluids. Two factors control the thermal conductivity of sedimentary rocks: the mineral composition (which is very variable) and porosity (which can be quite high). In addition the pore space can be filled by either air, water, or hydrocarbons. The conductivity of water is relatively low and varies with temperature as

$$\lambda_w = 0.93\left[1 + \left(\frac{T}{273}\right)\right]^{3/2} - 0.37\left[1 + \left(\frac{T}{273}\right)\right]^{5/2}$$

in the range 0°C to 400°C. The units are in W m^{-1}°C^{-1} and T is expressed in °C.

There is a large contrast between the conductivities of water and quartz at 0°C: $0.56/7.69$. Thus the presence of water reduces rock conductivity. If the pore fluid is air, the decrease in conductivity is even greater due to the very low conductivity of air.

$$\lambda_g = 0.024 \text{ W m}^{-1}°C^{-1}.$$

For these reasons near-surface sediments with large porosities generally have low conductivities, which increase noticeably as the rock compacts during burial.

If fractures are filled with air, then radiative transfer across fractures becomes important at much lower temperatures. Generally, convective transfer of air will be insignificant in vertical fractures whose widths are several millimeters or less. In horizontal fractures, convection becomes insignificant at much larger widths. One can estimate the heat flux J_r across a fracture by considering a fracture as two parallel plates whose extent is much larger than their separation d:

$$J_r = \frac{4e}{2-e}\sigma T^3(\Delta T) \equiv \lambda_r \frac{\Delta T}{d}.$$

Here $T = (T_1 + T_2)/2$, $\Delta T = T_1 - T_2$, σ is the Stefan-Boltzmann constant and e the rock emissivity. This relation is applicable when $T \gg \Delta T$. The radiative component is $\lambda_r = \frac{4e}{2-e}\sigma T^3 d$ and increases as T^3. The total conductivity of the fracture is the sum of the radiative transfer and the conductive component due to the gas molecules λ_g:

$$\lambda_{\text{fract}} = \lambda_g + \lambda_r.$$

Table X.3. THERMAL CONDUCTIVITY OF AN AIR-FILLED FRACTURE

Temperature $T(°C)$	0	38	93
λ_g(W m^{-1}°K^{-1})	0.0242	0.0266	0.0301
λ_r(W m^{-1}°K^{-1})	0.00376	0.00557	0.00908
λ_{fract}(W m^{-1}°K^{-1})	0.0280	0.0322	0.0392

Source: After Chan, T., and J. A. Jeffrey, 1983, *Scale and Water-Saturation Effects For Thermal Properties of Low-Porosity Rock*, Proceedings of the 24th U.S. Symposium on Rock Mechanics, Texas A & M University, Association of Engineering Geologists.

Table X.3 gives the respective values of λ_g and λ_r: λ_r increases with T and the effect of air-filled fractures in limiting heat transfer decreases with increasing temperature.

Sedimentary Rocks

The thermal conductivity of sedimentary rocks depends on their composition (mineralogy, volume fractions) and on microstructure (porosity). It varies with depth due to compaction and temperature. The influence of these effects can be illustrated by considering a well-sorted sandstone of porosity 0.45, saturated with water (at 0°C, 1 atm). Quartz (70%) and feldspar (30%) form the solid matrix. Assume further that the goethermal gradient is linear (30°C/km) and that the porosity decreases with depth according to the empirical law

$$\phi = \phi_o \exp(-z/14), \qquad \phi_o = 0.45.$$

The required parameters of the system considered are presented in table X.4.

The laws describing how the conductivities of quartz and feldspar vary with temperature in the range 0°–300°C are

$$\lambda_q = 7.67[1 + 4 \cdot 10^{-3}T]^{-1} \text{ W m}^{-1}°C^{-1}$$

$$\lambda_f = 2.30[1 + 5 \cdot 10^{-4}T]^{-1} \text{ W m}^{-1}°C^{-1}$$

where T is expressed in °C. The H.S. bounds are shown in figure X.11: the decrease in porosity produces an increase in λ, but this effect is

Table X.4. SANDSTONE

Component	Volume Fraction
Quartz	.70(1 − ϕ)
Feldspar (plagioclase)	.30(1 − ϕ)
Water	ϕ

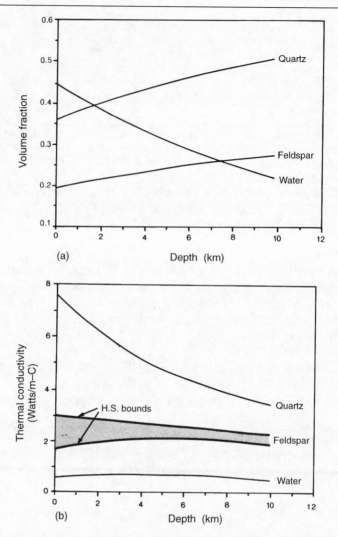

Fig. X.11. Thermal conductivity of sandstone as a function of depth. (*a*) Variation of composition due to compaction. (*b*) Variation of mineral conductivities, water, and sandstone (H.S. bounds).

compensated by a decrease of quartz conductivity at high temperatures. The total variation is quite small and the H.S. bounds restrict the conductivity to a narrow interval below 5 km in depth.

As for sandstone, compaction noticeably modifies the microstructure of clays and shales. Their porosity, which is initially very high, decreases rapidly with burial. To evaluate this effect assume that ϕ decreases with

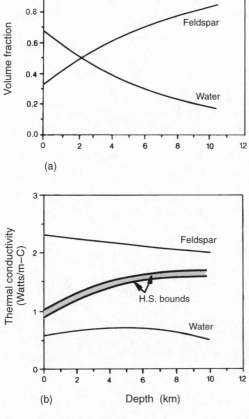

(a)

(b) Depth (km)

Fig. X.12. Thermal conductivity of an average shale as a function of depth. (*a*) Variation of composition due to compaction. (*b*) Variation of mineral conductivities, water, and shale (H.S. bounds).

z according to the empirical law:

$$\phi = \phi_o \exp(-z/7), \qquad \phi_o = 0.70.$$

The mineral conductivity of clays is approximately that of amorphous silica. Because silicate glass is in the same range as plagioclase feldspar and exhibits a similar temperature dependence, we will use the expression for feldspar to approximate the mineral conductivity of clays. With these assumptions, the computed shale conductivity rises from 0.92 W $m^{-1}°C^{-1}$ at the surface to 1.55 W $m^{-1}°C^{-1}$ at 10 km, where the porosity has been reduced to 0.17. The H.S. bounds again define a narrow range of values for the effective conductivity (fig. X.12).

The previous examples are meant to be for illustrative purposes only, with the focus being on the methodology of estimating $\lambda(z)$ from its mineral components and microstructure. In particular the laws utilized for the variation $\phi(z)$ are gross approximations that minimize the variation of ϕ. These examples also show that it is possible to estimate the effective thermal conductivity within a very small range of uncertainty.

PROBLEMS

X.a. Convective and Conductive Fluxes

Consider a porous medium for which l is the characteristic distance of fluid transport. The expressions for the conductive and convective heat fluxes are $\lambda \dfrac{\Delta T}{\Delta x}$ and $\rho c q \, \Delta T$, respectively, and ΔT is the temperature difference over the distance $\Delta x = l$.

(1) Determine the domain (q, l) where the conductive flux is dominant.
(2) Discuss this domain in terms of the permeability and fluid pressure gradient, assuming that the fluid is water.

X.b. Radiative Conductivity

According to Planck's theory, the radiative conductivity is due to a "gas" of photons. It is given by the expression $\lambda_r = \dfrac{16}{3} \dfrac{n^2}{\alpha} \sigma T^3$ where n is the index of refraction of the medium, α^{-1} the mean free path of the photons, and σ the Stefan-Boltzmann constant ($\sigma = 5.6 \times 10^{-5}$ ergs cm^{-1} s^{-1}°K^{-4}). Discuss the temperature domain where λ_r becomes dominant. (Assume that α^{-1} is in the range 1–10 cm.)

X.c. Effective Conductivities

(1) Using the values in table X.4, calculate the effective conductivities corresponding to (1) the harmonic average; (2) the arithmetic average; (3) the geometric average.
(2) Compare these predictions to the H.S. bounds.

X.d. Conducting Sphere

Consider a homogeneous medium of conductivity λ_o, in which there flows a uniform heat flux $\vec{J}_o$. A homogeneous sphere of conductivity λ and radius R is then introduced into this medium.

(1) Show that the temperature is given by:

$$T = -\left(\vec{J}_o \cdot \vec{r}\right)\lambda_o^{-1} + A\frac{\left(\vec{J}_o \cdot \vec{r}\right)}{r^3} \qquad \text{exterior to the sphere} \quad (r > R)$$

$$T = -B\left(\vec{J}_o \cdot \vec{r}\right) \qquad \text{interior to the sphere} \quad (r < R).$$

(2) Determine the constants A and B by utilizing the appropriate boundary conditions at $r = R$.

(3) For an ensemble of spheres (neglecting their interactions), calculate the quantities: $\langle \vec{\nabla} T \rangle, \langle \vec{J} \rangle$. and thereby deduce λ_{eff}.

REFERENCES

Berman, R., F. E. Simon, P. G. Klemens, and T. M. Fry. 1950. *Nature* 166:864.

Birch, F., and H. Clark. 1940a. The Thermal Conductivity of Rocks and Its Dependence Upon Temperature and Composition, 1. *American Journal of Science* 238:529–58.

———. 1940b. The Thermal Conductivity of Rocks and Its Dependence Upon Temperature and Composition 2. *American Journal of Science* 238:613–35.

Cermak, V., and L. Rybach. 1982. Thermal Conductivity and Specific Heat of Minerals and Rocks, in *Physical Properties of Rocks*, vol. 1, Landolt-Bornstein-New Series. Berlin, Heidelberg: Springer-Verlag.

Robertson, E. C. 1988. *Thermal Properties of Rocks*. Open-File Report 88-441, United States Department of the Interior Geologic Survey, Reston, Virginia.

Roy, R. F., A. E. Beck, and Y. S. Touloukian. 1981. Thermophysical Properties of Rocks, in *Physical Properties of Rocks and Minerals*, McGraw-Hill/Cindas Data Series on Material Properties, vol. II-2, eds. Y. S. Toulokian and C. Y. Ho, pp. 409–502.

XI. Magnetic Properties

THE TOTAL static magnetic field, as measured at a given point on the earth's surface, can be analyzed as the superposition of two distinct fields: the earth's global magnetic field and a local field caused by rock magnetization. Here we will not consider transient fields although they are of fundamental importance for investigating the earth's electrical conductivity structure. The local static field is due to an induced rock magnetization which is proportional to the applied field and a frozen-in magnetization called remnant magnetization, which exists even when the applied field is zero. This remnant magnetization is the magnetic memory of a rock. The abundance of iron-rich minerals such as magnetite plays a very important role in determining the strength of the local field. Thus measurements of magnetic susceptibility are very useful in the prospecting of metallic ores. However, the most spectacular application of rock magnetism has been in fundamental research. The existence of a remnant magnetization in rocks has played a key role in the development of the theory of plate tectonics, by making it possible to measure quantitatively the ocean spreading rates.

MINERAL MAGNETISM

The magnetic properties of rocks are due to a volumetrically very small fraction of magnetic minerals. These minerals do not need to be organized in a connected network, so that percolation effects play no role. The magnetic properties of rocks directly reflect the magnetic properties of these minerals which are the "carriers" of magnetization.

Magnetic Quantities

In a vacuum, the relation between magnetic induction $\vec{B}$ and the magnetic field $\vec{H}$ is, in S.I. units:

$$\vec{B} = \mu_o \vec{H} \qquad \text{XI.1}$$

where μ_o is the permeability of a vacuum: $\mu_o = 4\pi \, 10^{-7} = 1.257 \times 10^{-6}$ Hm^{-1}. The units of $\vec{H}$ and $\vec{B}$ are respectively the Ampere/meter and Tesla. When the intensity of spontaneous magnetization is zero, the above equation becomes (for an isotropic medium):

$$\vec{B} = \mu \vec{H}. \qquad \text{XI.2}$$

Here μ is the magnetic permeability of the medium.

The magnetization $\vec{M}$ is defined as the magnetic moment per unit volume. Within the linear approximation, induced magnetization is proportional to the field $\vec{H}$:

$$\vec{M} = \chi \vec{H} \qquad\qquad \text{XI.3}$$

χ is a dimensionless constant called the magnetic susceptibility. Equation XI.2 is obtained by combining XI.3 and the relation which defines $\vec{M}$:

$$\vec{B} = \mu_o\left(\vec{H} + \vec{M}\right). \qquad\qquad \text{XI.4}$$

The previous equations imply that $\mu = \mu_o(1 + \chi)$. Generally $\chi \ll 1$ for rocks, so that $\mu \approx \mu_o$. Thus rocks cannot be identified on the basis of their magnetic permeability values. They can be identified however from their magnetic susceptibility values, which vary over several orders of magnitude. χ is thus the important parameter.

The c.g.s. system of units is commonly used in magnetism. In these units equations XI.4 and XI.2 become:

$$\vec{B} = \vec{H} + 4\pi\vec{M} = (1 + 4\pi\chi)\vec{H}.$$

The c.g.s. unit for B is 1 Gauss = 10^{-4} Tesla, and for $\vec{H}$, it is 1 Oersted = 10^2 Am^{-1}. The correspondence between units is as follows:

$$\frac{\mu}{\mu_o}(\text{S.I.}) = 1 + \chi(\text{S.I.}) = 1 + 4\pi\chi(\text{c.g.s.})$$

Paramagnetism and Ferromagnetism

Most minerals have a very low magnetic susceptibility, lower than 10^{-4}. This means that the induced magnetization is negligible and these minerals will not produce any magnetic anomaly. In order to understand the magnetic properties of solids, it is necessary to look at the atomic scale. Each atom carries a magnetic moment which is primarily due to the electron spin and electron orbital angular moment. The atomic magnetic moment $\vec{\mu}$ is a quantum mechanical parameter. When an induction $\vec{B}$ is applied, $\vec{\mu}$ can take on several possible values which correspond to a discrete set of energy levels $E_i = -\mu_i B$. Magnetic induction tends to orient the moments parallel to each other. Thermal agitation, on the other hand, tends to randomize this order. For a population of identical atoms and energy levels $E_1 \cdots E_k$, the number of atoms $n(E_i)$ per level per unit volume, at a temperature T, is given by Maxwell-Boltzmann distribution:

$$\frac{n(E_i)}{N} = \frac{\exp - (E_i/kT)}{\displaystyle\sum_{i=1}^{k} \exp - (E_i/kT)}$$

where N is the total number of atoms per unit volume and k is the Boltzmann constant ($k = 1.38 \times 10^{-23}$ J K^{-1}). The total magnetic moment of a solid per unit volume is:

$$M = \sum_{i=1}^{N} n_i \mu_i.$$

In the simple case where there exist only two energy levels, that is, if $\mu = \pm\mu_b$:

$$E_1 = \mu_b B \quad \text{and} \quad E_2 = -\mu_b B.$$

Letting $x = \mu_b B/kT$, one then finds:

$$\frac{n_1}{N} = \frac{e^{-x}}{e^x + e^{-x}}, \frac{n_2}{N} = \frac{e^x}{e^x + e^{-x}}, \quad \text{and} \quad M = (n_2 - n_1)\mu_b$$

with the final result

$$M = \left(\frac{e^x - e^{-x}}{e^x + e^{-x}}\right) N\mu_b.$$

If μ_b is Bohr's magneton (the magnetic moment associated with a spin 1/2 electron), then $\mu_b = eh/4\pi m = 9.27 \times 10^{-24}$ J T^{-1}, where e and m are respectively the charge and mass of an electron and h is Planck's constant. It is then easy to show that $\chi \ll 1$, except when $T \to 0°$K or when B reaches very high values (the thermal energy kT is usually much larger than the magnetic energy $\mu_b B$).

At the earth's surface, the induction B is on the order of 1 Gauss (10^{-4} Tesla). When $B = 1$ Gauss and $T = 10^3$ °K, $x \approx 10^{-7}$. When $x \ll 1$, the previous equation can be simplified by using the approximation $e^x \approx 1 + x$. One then finds

$$M \approx N\mu_b x = \frac{N\mu_b^2 B}{kT}.$$

With $B \approx \mu_o H$ and $M = \chi H$, we arrive at an expression for χ:

$$\chi = \frac{N\mu_b^2 \mu_o}{kT}. \qquad \text{XI.5}$$

Equation XI.5 shows that the susceptibility varies inversely to temperature ($\chi \sim 1/T$): this is the Curie law. Assuming that there is one magnetic moment in a cube of edge 10^{-9} m, that is, $N = 10^{27}$ m^{-3}, $\chi \approx 10^{-4} - 10^{-5}$ in the temperature range $T = 100-1000°$K. A graph showing how M varies with H is presented in figure XI.1. The maximum value of M, the "saturation magnetization" M_s, is reached as $T \to 0$ or $B \to \infty$. This "saturation magnetization" is $M_s = N\mu_b$.

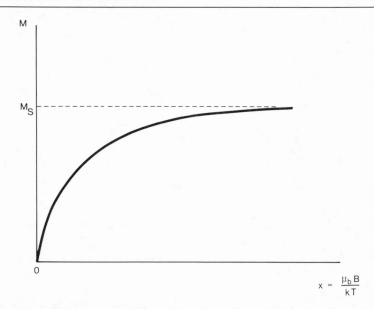

Fig. XI.1. Paramagnetic mineral: variation of magnetization M as a function of induction B or reciprocal temperature T^{-1}.

The low M values observed for paramagnetic minerals result from the competition between two effects: the applied field which tends to order the moments parallel to the field, and thermal agitation, which tends to randomize this order. As we have seen, the second effect dominates over the first. The existence of high magnetization and high susceptibility values for some minerals results from a different interaction, the exchange interaction, which tends to line up magnetic moments parallel to each other (at least at low temperatures below a critical value). The exchange interaction is a quantum mechanical effect, related to the overlap of charge distributions of atoms and to the Pauli principle, which says that two electrons of the same spin cannot be at the same place at the same time. The exchange interaction is not a dipole-dipole interaction but an electrostatic interaction, and this is why the exchange energy is quite high. Below a critical temperature (the Curie temperature T_c), the exchange interaction dominates and the magnetic moments are oriented as a result of their strong mutual interaction. This is the ferromagnetic state with a non-zero spontaneous magnetization. The mineral is a permanent magnet. At $T = T_c$, there is a phase transition from the ordered ferromagnetic phase to the disordered paramagnetic phase. Above T_c, the mineral has no spontaneous magnetization. An exact theory requires a complete description of

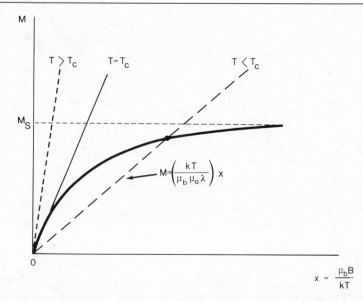

Fig. XI.2. Mean field approximation in ferromagnetism: M is obtained as the intersection of the paramagnetic curve and the straight line $M = \dfrac{kT}{\mu_b \mu_o \lambda} x$.

interactions between moments. In order to avoid this complexity, P. Weiss imagined that each moment experiences an effective mean field $H_m = \lambda M$, where M represents the average magnetization of all the other moments. This is a self-consistent approximation, similar to that presented in chapter III for calculating effective medium properties. Then:

$$ M = \left(\frac{e^x - e^{-x}}{e^x + e^{-x}} \right) N\mu_b \quad \text{and} \quad x = \frac{\mu_b B}{kT} = \frac{\mu_b \mu_o \lambda M}{kT} $$

because $B = \mu_o H_m = \mu_o \lambda M$. The last equation is of the self-consistent type: $M = f(M)$. The solution of this equation is illustrated graphically in figure XI.2. For T above T_c there is no solution except for $M = 0$. When $T < T_c$, there is a non-zero solution for M: the mineral is ferromagnetic and has a permanent magnetization.

In the paramagnetic phase, $T > T_c$:

$$ \vec{M} = \chi \vec{H}, \qquad \vec{H} = \vec{H}_o + \vec{H}_m, \qquad \chi = \frac{C}{T} \text{ (Curie Law)} $$

the field $\vec{H}$ is the sum of the applied field $\vec{H}_o$ and the average field $\vec{H}_m$ due to the other moments (Weiss field). Assuming that $\vec{H}_m = \lambda \vec{M}$, one then finds

$$\vec{M} = \left(\frac{C}{T - T_c} \right) \vec{H}_o \quad (T > T_c) \qquad \qquad \text{XI.6}$$

where $T_c = \lambda C$ is the Curie temperature.

The exchange energy can be either positive or negative. In the first case, moments will orient themselves parallel to each other and the interaction is ferromagnetic. In the second case, moments will orient themselves antiparallel to each other and the interaction is antiferromagnetic. An important example is that of magnetite Fe_3O_4. This mineral has a spinel crystal structure which has 24 cations in a unit cube:

$$8 \; Fe^{3+} \; \text{occupy tetrahedral or } A \text{ sites;}$$

$$8 \; Fe^{3+} \; \text{occupy octahedral or } B \text{ sites;}$$

$$8 \; Fe^{2+} \; \text{occupy octahedral or } B \text{ sites.}$$

Three types of exchange interactions have to be considered: $A - A$, $B - B$, and $A - B$. The strongest one is the $A - B$ interaction, which is antiferromagnetic, so that the A spins are parallel to each other, the B spins are parallel to each other, but the A spins are antiparallel to the B spins. The resulting magnetic order is an example of ferromagnetic order. The saturation magnetization is 4 μ_b per molecule, which corresponds to the moment of Fe^{2+} cations since Fe^{3+} moments are antiparallel to each other.

Magnetic Domains

When $T \ll T_c$, magnetic moments are all parallel (or antiparallel) in a ferromagnetic mineral, at least at the microscopic scale (~ 10 nm). However, if no external field is applied, the total magnetization is very low compared to the expected value $M_s = N\mu_b$. Only when an external field is applied is M observed to increase up to M_s. This behavior is observed for single crystals as well as aggregates. When viewed under closer scrutiny, one sees that the actual samples are composed of small regions called "domains." In each domain, the magnetization value is M_s, but the directions of magnetization of the domains are not aligned. The overall macroscopic magnetization is thus lower than M_s and can even be close to zero (fig. XI.3).

Under the action of an applied magnetic field, the domains which are favorably oriented grow at the expense of the other domains. If the

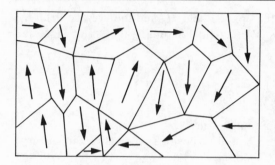

Fig. XI.3. Domains in a ferromagnetic crystal. Inside each domain, moments are aligned parallel to each other. Domain size is typically $10^{-2} - 1\ \mu\text{m}$.

applied field is sufficiently strong, there can be a rotation of the local magnetization vectors toward the direction of the field. The overall magnetization M of the ferromagnetic solid varies with the applied field H_o. Figure XI.4 shows such a representative magnetization curve. When the field is decreased, the magnetization follows a different path: a hysteresis loop is defined (fig. XI.5). The remnant magnetization is M_r, which is the value of M for $H_o = 0$. The coercive field H_c is the required reverse field to bring M to zero. If the value of H_c is high, the mineral has a good magnetic "memory." For instance, $M_c \approx 10^4\ \text{Am}^{-1}$ for magnetite and $10^5\ \text{Am}^{-1}$ for hematite.

From a macroscopic point of view, ferromagnetic solids can be considered to be "pseudo-paramagnetic." This is because magnetization M varies with H_o and the susceptibility χ is calculated from $M = \chi H_o$.

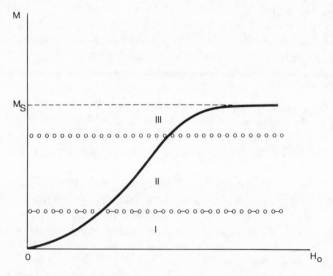

Fig. XI.4. First magnetization curve for a ferromagnet: I—reversible displacements of domain walls; II—irreversible displacements; III—domain rotations.

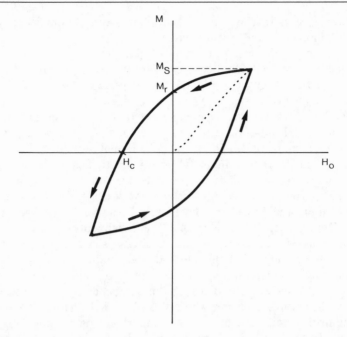

Fig. XI.5. Hysteresis loop of a ferromagnet. M_r is the remnant magnetization and H_c the coercive field (dotted line is the first magnetization curve).

But in this case, the susceptibility values are very large compared to those of true paramagnetic crystals. An approximate value of χ can be obtained by utilizing the paramagnetic theory discussed previously and substituting domains for atomic moments. The domains play a similar role in ferromagnetism as the atomic moments do in paramagnetism.

Assuming that all domains are identical and contain n moments, substitute $n\mu_b$ for μ_b. The following substitutions can then be carried out:

N, number of moments per unit volume

$$\to \frac{N}{n}, \text{ number of domains per unit volume}$$

μ_b, atomic moment $\to n\mu_b$, domain moment

$\chi_p = N(\mu_b)^2 \dfrac{\mu_o}{kT}$, paramagnetic susceptibility

$$\to \chi_f = \frac{N}{n}(n\mu_b)^2 \frac{\mu_o}{kT}, \text{ ferromagnetic susceptibility.}$$

It is easily seen that $\chi_f = n\chi_p$. Since an average domain contains 10^4–10^6 moments, the pseudo-susceptibility χ_f of a ferromagnetic mineral is very high, close to unity or even larger.

Magnetic domains exist because their organization minimizes the total magnetic energy of a mineral. As suggested in figure XI.3, moments of neighboring domains form closed magnetic loops. These loops "capture" or localize the magnetic field (due to the magnetization) inside the solid. If all of the domains shown in figure XI.3 were parallel, the resulting moment would be high and there would exist a large magnetic field external to the solid (the magnetic field of the magnet). The existence of several domains prevents this and lowers the magnetic energy. But there is an energy cost for introducing a discontinuity in magnetization, the so-called "Bloch wall," between two adjacent domains. Domain size is thus controlled by a competition between these two opposing effects. Below a critical size, a single crystal is always magnetized as a single domain. Single-domain grains can be oriented by rotation only, but this process may require large fields. Such materials will exhibit large coercive fields. A similar behavior is observed for grains with multiple domains, where the Bloch walls have a low mobility. Impurities and defects play an important role in lowering the mobility of Bloch walls. Some grains are called "pseudo single domain" because they behave as single-domain grains. These grains are in fact organized into several domains but most of the resulting magnetization is due to walls close to the grain boundary.

Domains constitute the internal microstructure of grains. Contrary to what we have seen for most of the other rock physical properties, this internal grain structure plays a major role in the magnetic properties of rocks.

Magnetic Minerals

The important magnetic minerals are iron-rich minerals. Fe^{2+} and Fe^{3+} cations carry magnetic moments because of the incompletely filled 3d electron energy band. The spin of the 3d electrons for these cations is as follows:

$$Fe^{2+} \quad 3d^6 \quad \uparrow\downarrow \quad \uparrow \quad \uparrow \quad \uparrow \quad \uparrow \quad \text{magnetic moment } 4\mu_b$$
$$Fe^{3+} \quad 3d^5 \quad \uparrow \quad \uparrow \quad \uparrow \quad \uparrow \quad \uparrow \quad \text{magnetic moment } 5\mu_b$$

Iron oxides Fe_3O_4 and Fe_2O_3 are the most common magnetic minerals. They are end members of solid solutions with titanium oxides. The Ti^{4+} cation does not carry any magnetic moment but its radius (0.68 Å) is intermediate between Fe^{3+} (radius 0.64 Å) and Fe^{2+} (radius 0.74 Å).

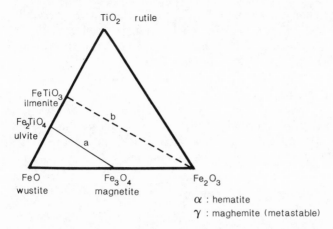

Fig. XI.6. The principal magnetic minerals. The solid line corresponds to titanomagnetites while the dotted line corresponds to titanohematites and titanomaghemites.

The principal magnetic minerals can be classified according to the triangular diagram shown in figure XI.6. The three corners of this diagram represent TiO_2, FeO, Fe_2O_3. *Titanomagnetites* are found in intrusive rocks (basalt) and have a formula of the type $Fe_{3-x}Ti_xO_4$. The atomic magnetic structure for the end member magnetite has been presented above. Magnetite has a high value for the saturation magnetization M_s (4.8×10^5 Am^2/m^3 or 93 Am^2/kg). M_s decreases, as does the Curie temperature T_c, when the Fe^{3+} content increases (for magnetite $T_c = 580°C$). The coercive field H_c depends on grain size. Its maximum value is 3×10^4 Am^{-1}, but lower values are commonly observed.

Titanohematites $Fe_{2-x}Ti_xO_3$ are weakly antiferromagnetic minerals that are found in both intrusive and sedimentary rocks. These minerals have lower saturation magnetization values than do the titanomagnetites: M_s is approximately 200 times less for hematite than for magnetite (i.e., 2.2×10^3 Am^2/m^3 or 0.5 Am^2/kg). Both M_s and T_c decrease as the iron content increases (for hematite $T_c = 680°C$). The hematite-ilmenite solid solution is complete only at $T > 700°C$. Slow cooling of this solid solution leads to exsolution of ilmenite- and hematite-rich crystals. An important characteristic property of titanohematites is their high H_c value (about 3×10^5 Am^{-1}), about ten times higher than the H_c value measured in titanomagnetites.

Titanomaghemites are oxidized titanomagnetites. They are non-stochiometric and metastable minerals. Remnant magnetization found in basalt is primarily due to these minerals.

There exist other magnetic minerals. For example, pyrrothite (Fe_7S_8-FeS) is found in basic igneous rocks, in high-temperature hydrothermal veins, and in sediments. Goethite (αFeO-OH) occurs commonly as a weathering product of iron-bearing minerals.

ROCK MAGNETIZATION

Rock magnetization results from two contributions: the induced magnetization M_i and remnant magnetization M_r. The Koenigsberger ratio is defined as the ratio of these two quantities:

$$Q = \frac{M_r}{M_i}. \qquad \qquad \text{XI.7}$$

For most common rocks, Q varies in the range 0–100 (table XI.1).

Induced Magnetization

The importance of induced magnetization depends on the rock's magnetic susceptibility, which in turn depends on the magnetic minerals that it contains. Sedimentary rocks contain very small amounts of magnetic minerals and thus have a very low magnetic susceptibility (table XI.2). Basic rocks exhibit much higher susceptibility values. To a first approximation, the rock susceptibility χ depends on the magnetite concentration c and magnetite susceptibility χ_m:

$$\chi \approx c \chi_m.$$

Because χ_m is close to unity, the magnetic interaction between grains cannot be neglected when c is very high. Each grain opposes the

Table XI.1. KOENIGSBERGER RATIO

Rock	Acidic Intrusive Veins	Basic Intrusive Veins	Basaltic Lava
Q	0–1	1–10	100

Source: Strangway, D. W., 1967, Magnetic Characteristics of Rocks, in *Mining Geophysics* II. Soc. Expl. Geophys. Tulsa, Okla.

Table XI.2. MAGNETIC SUSCEPTIBILITIES OF ROCKS

Rock Type	Sedimentary	Granites and Gneisses	Intrusive Basic Rocks
χ	$< 10^{-4}$	10^{-4}–10^{-3}	$> 10^{-3}$

applied field H_o by producing a demagnetizing field. The local field interior to the rock sample is then

$$H = H_o - \alpha M$$

where α is the demagnetization factor (close to unity) and depends on the geometry of the solid. Thus $M = \chi_m H = \chi_m (H_o - \alpha M)$, and the apparent susceptibility $\chi_a = M/H_o$ is:

$$\chi_a = \frac{\chi_m}{1 + \alpha \chi_m}.$$

If $\alpha \chi_m \gg 1$, $\chi_a \approx 1/\alpha$: the apparent susceptibility is controlled by the demagnetizing field, and χ_a does not depend on χ_m. Usually the concentration of magnetic minerals is small, so that $c\chi_m$ must be substituted for χ_m. Consequently:

$$\chi_a = \frac{c\chi_m}{1 + \alpha c \chi_m}. \qquad \text{XI.8}$$

Usually $\alpha c \chi_m \ll 1$, so that $\chi_a \approx c \chi_m$. In some cases however, $c \approx 10^{-2}$ and $\chi_m \approx 10^2$, so that $\alpha c \chi_m$ cannot be neglected as compared to unity. The value of α depends on the geometry of the magnetic grains.

Remnant Magnetization

The existence of a remnant magnetization is a specific property of ferromagnetic minerals. Single-domain grains and pseudo single-domain grains are the best remnant magnetization carriers, because their coercive field H_c is high. Remnant magnetization can result from various processes, with thermo-remnant magnetization (TRM) being the principal one. TRM is created when a ferromagnetic mineral is cooled below its Curie temperature in a non-zero field. The magnetization, being parallel to the existing field at the time of cooling, is "frozen" in the solid. For example, consider a population of single-domain grains. Assume further that all grains have the same moment $\vec{m}$, which can be either parallel or antiparallel to $\vec{B}$. Utilizing the previous calculation for atomic moments, we can derive:

$$\vec{M} = N\vec{m} \tanh\left(\frac{mB}{kT}\right). \qquad \text{XI.9}$$

N is the number of single-domain grains per unit volume. The above equilibrium result is only valid if the kinetics of grain reversal is fast enough, which is true at high temperatures. Here $\vec{m}$ represents the moment of a grain, that is, of a domain, because all grains are assumed to be comprised of a single domain. The exact value of m depends on

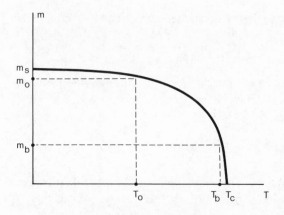

Fig. XI.7. Temperature dependence of domain magnetization. This graph is derived from fig. XI.2.

temperature T as shown in figure XI.7. This m-T plot is derived from figure XI.2. When T decreases below T_c, m varies between 0 and m_s. The mobility of domains decreases as the sample cools, and at a temperature $T = T_b$, spontaneous reversal of grains becomes almost impossible. T_b is called the blocking temperature and is defined as the temperature at which the time constant τ for grain reversal is $\tau = 10^3$ s. If T decreases down to 300°C, τ values become extremely long and the domains can be considered completely "frozen" in. We can then utilize equation XI.9 with $T = T_b$. Therefore at a temperature T_o, the magnetization $\vec{M}$ is:

$$\vec{M} = n\vec{m}_o \tanh\left(\frac{m_o B}{kT_b}\right).$$

As indicated in figure XI.7, $m_o \gg m_b$, and the magnetization $\vec{M}$ can reach very high values.

Remnant magnetization may also be the result of other processes such as:

(a) viscous remnant magnetization (*VRM*): this is the magnetization developed in a constant field. *VRM* is not negligible when domains are not very stable;

(b) isothermal remnant magnetization (*IRM*): this is the magnetization developed at constant temperature in a very strong field. *IRM* is negligible in general, except for rocks struck by lightning;

(c) chemical remnant magnetization (*CRM*), which results from the growth of magnetic crystals (Fe_3O_4 oxidized in Fe_2O_3, for instance). *CRM* can be acquired at low temperatures;

(d) detrital remnant magnetization (*DRM*), which is the result of sediment deposition.

Depending on the rock type, natural remnant magnetization (NRM) is dominated by one or the other of these processes. For volcanic rocks, one finds:

$$NRM = TRM + CRM + VRM.$$

The TRM component is acquired by cooling down to $T_b \approx T_c - 50°C$. The CRM component is mainly due to exsolution, while the VRM component is the result of the evolution of some rather unstable domains. CRM can also be a secondary magnetization superimposed on the original or primary magnetization, when ferrous minerals are oxidized. For sedimentary rocks, a different combination holds:

$$NRM = CRM + VRM + DRM.$$

Utilizing NRM in sedimentary rocks is difficult, because NRM values are very low and the processes complex.

PROBLEMS

XI.a. Ferromagnetism

(1) Calculate the constant λ (Weiss field $H_m = \lambda M$) for magnetite using $T_c = 580°C$.

(2) Derive the value of H_m and an estimate of the exchange energy. Compare this result with the magnetic energy of a moment in an applied field of 10^{-4} Tesla (earth's field).

XI.b. Ferromagnetic Domains

Consider a thin planar layer of magnetic mineral (thickness L_z, length L_x, and width L_y; L_x and $L_y \gg L_z$. This layer is divided into (L_x/d) domains of width d. Assume that the demagnetizing field is $H \approx M$ at distances less than d of the lamella faces and zero further. The wall surface energy is σ.

(1) Show that the total magnetic energy is $U \approx \sigma L_x L_y L_z/d + \mu_o M^2 L_x L_y d$.

(2) Derive d as a function of M, σ, and L_z. Use $\sigma = 10^{-3}$ J m^{-2}, $L_z = 1$ mm, $M = 10^5$ Am^{-1}.

XI.c. Stability of Remnant Magnetization

Consider a population of N single-domain grains. At $t = 0$, H is decreased to zero. Grains can reverse with a reversal probability (frequency) $\dfrac{dP}{dt} = \nu = \nu_o \exp - (E/kT)$; $E \approx \mu_o MVH_c$, where V is the volume of a grain and $\nu_o = 10^8$ s^{-1}. The relaxation time is defined as $\tau = \nu^{-1}$.

(1) Compute a numerical estimate for E.

(2) Show that $M = M_o e^{-2t/\tau}$.

(3) Compute τ at $T = 300°K$ with the assumption that at $T = T_b = 800°K$, $\tau(T_b) = 10^3 \ s$.

REFERENCES

Deer, W. A., R. A. Howie, and J. Zussman. 1980. *An Introduction to Rock Forming Minerals*. London: Longman.

Stacey, F. D., and S. K. Banerjee. 1974. The physical principles of rock magnetism. *Developments in Solid Earth Geophysics* 5. Amsterdam: Elsevier.

Westphal, M. 1986. *Paléomagnétisme et magnétisme des roches*. Paris: Doin.

Index